THE KINETICS OF MUSCLE CONTRACTION

PERGAMON STUDIES IN THE LIFE SCIENCES

WITH A FOREWORD BY D. NOBLE
BALLIOL COLLEGE, OXFORD

THE KINETICS OF MUSCLE CONTRACTION

D. C. S. WHITE

Department of Biology, University of York, Heslington, York

and

JOHN THORSON

Department of Zoology, Oxford

PERGAMON PRESS

Oxford · New York · Toronto · Sydney · Braunschweig

Pergamon Press Ltd., Headington Hill Hall, Oxford
Pergamon Press Inc., Maxwell House, Fairview Park, Elmsford, New York 10523
Pergamon of Canada Ltd., 207 Queen's Quay West, Toronto 1
Pergamon Press (Aust.) Pty. Ltd., 19a Boundary Street, Rushcutters Bay, N.S.W. 2011, Australia
Pergamon Press GmbH, Burgplatz 1, Braunschweig 3300, West Germany

---

First edition 1975

Originally published in
Progress in Biophysics and Molecular Biology
(Editors: J. A. V. Butler and D. Noble),
Volume 27, 1973. Reprinted here with slight revision.

Library of Congress Cataloging in Publication Data

White, David Clifford Stephen.
The kinetics of muscle contraction.

(Pergamon studies in the life sciences)
"Originally published in Progress in biophysics and molecular biology . . . volume 27, 1973."
Bibliography: p.
Includes index.
1. Muscle contraction. I. Thorson, John, joint author. II. Title. [DNLM: 1. Kinetics. 2. Muscle contraction. WE545 W583k]
QP321.W47 1974 612'.741 74-20860
ISBN 0-08-018149-X lib. bdg.

Printed in Great Britain by Biddles Ltd., Guildford, Surrey

CONTENTS

FOREWORD

THE study of the mechanism of muscle contraction has advanced very rapidly in the last few years and it has now become possible to perform experiments that determine the kinetics of the protein cross-bridge reactions that are responsible for generating mechanical force. Muscle biophysics has always been a popular subject in student courses and the excitement that these advances have produced has further stimulated the teaching of the subject. Drs. White and Thorson have produced an excellent explanatory review of the kinetics of muscle contraction at a very opportune time. Their review appeared originally in *Progress in Biophysics and Molecular Biology* (volume 27) and it is evident from the well-thumbed pages of the copy of this volume in my own university's science library that this review is already very popular with undergraduate students. To increase its availability we have therefore arranged its publication as a separate monograph. The authors have added some additional notes (Appendix III, p. 77) to reflect recent developments since the review originally appeared.

This is the first of a series of such publications and we shall take the opportunity to publish further articles from *Progress in Biophysics* where it is clear that the authors have succeeded in presenting their subject in a manner understandable to students and in a form that lends itself to publication as a monograph. The series may also include monographs written specifically for the series.

Balliol College, Oxford

DENIS NOBLE

I. INTRODUCTION

After two decades of steady growth, the understanding of muscle at the macromolecular level is showing signs of acceleration. Advances in rapid-reaction biochemical kinetics through such work as Taylor's and Trentham's on the one hand, coupled with improved temporal resolution of the mechanical dynamics by A. F. Huxley and his colleagues on the other, promise a new stage in the dialogue between the chemistry and the mechanics. Moreover, the current language of this dialogue, the theory of the cross-bridge cycle, is in an heuristic state as well, since experimentally distinguishable alternatives within A. F. Huxley's theory are becoming clear.

One price of such progress is that a survey article is difficult to write. Fortunately the 1972 Cold Spring Harbor Symposium, which coincides with this writing, has had the effect of synchronizing the reporting of ongoing work in the several disciplines concerned. The generosity of colleagues in sharing with us their manuscripts for that and other volumes has enabled a more current review than would otherwise have been feasible.

Our main aim is a critical review of the work which defines the present conceptual role of the cross-bridge cycle in the dynamics of contraction. Just two digressions will be apparent: first, we review separately certain interpretations of A. V. Hill's early model of muscle and go to some lengths to show just why these ceased to be either sufficient or heuristic with the advent of the sliding-filament theory. Second, and for obvious reasons, we give special attention to the current biochemical kinetics.

No attempt will be made to survey the entire literature; rather we shall try to explain the ideas and methodology underlying the development of a few key topics. We also include some new and informative calculations on the relationships of filament elasticity and the three-state cross-bridge cycle to recent results from both vertebrate and insect muscle.

There is considerable variety in the structure of muscle. For a number of reasons, vertebrate striated muscle has been studied most extensively. The muscles are of a convenient size for mechanical experiments, and judicious choice provides examples that are uniform in both geometry and fibre type. The frog sartorius has been used for most mechanical work on intact muscles, and the frog semitendinosus for most work on single fibres (largely because of the comparative ease of dissection of fibres from this muscle). Frogs are particularly suitable animals, being easy to keep and having muscles which work well at low temperatures. Striated muscle is much more regular in its structural features than other types; this results in advantages for study both in the electron microscope and by X-ray diffraction. For biochemical studies, vertebrate striated muscle has the particular advantage that it is available in very large quantities. Rabbit muscle is often used.

Although studies of such tissues as cardiac muscle, vertebrate smooth muscle, non-fibrillar insect skeletal muscle, and the "catch" muscle of molluscs pose critical comparative questions, we shall confine our attention to vertebrate striated and insect fibrillar flight muscles; most of the current theories of contraction are derived from their analysis. The peculiar advantages of the insect flight muscle for the study of the mechanism of contraction were realized in the 1950s by Pringle, and most of the work on this muscle has been carried out either by, or in collaboration with, his group at Oxford. Structurally, insect flight muscle resembles vertebrate striated muscle in many respects. Considerable impetus to this work came from the finding of sources of giant water-bugs of the genus *Lethocerus*, which is fairly widespread in tropical regions, and has muscles about 10–15 mm long. In this article, "insect flight muscle" will denote the dorsal longitudinal muscle from *Lethocerus* unless we say otherwise.

The first part of this article deals with the properties of muscle in the steady state, i.e. where conditions are held constant during the experiment. Such steady-state conditions are to be distinguished from the "transient" or kinetic conditions which occur when one of the controlled variables is changed. Thus if the length of the muscle is changed during an isometric contraction, or the load changed during an isotonic contraction, or the environment of the contractile proteins changed during any experiment, then the steady-state conditions will be perturbed. If the new conditions are then maintained at a constant level there will be a measurable transition to new steady-state conditions. The way in which the muscle responds during this transition is of considerable interest and will be the basis of the central part of this article. Obviously the perturbation can take any form. In practice the simplest is a sudden change from one constant level to another constant level. This we will call a step change in the particular parameter. However, other forms of perturbation are used. In mechanical experiments in particular, a sinusoidal perturbation of muscle length is often applied.

Research on muscle can be described under four main headings: structural, biochemical, mechanical, and energetic. There is considerable knowledge of the properties of muscle in each of these areas under steady-state conditions, but only recently has it been possible to measure the change from one steady state to another with sufficient time resolution in mechanical and biochemical experiments, and this stage has still not been reached in experiments investigating structural or energetic changes. One aim of this article is to relate, so far as is possible at the moment, the work that is being done in these four areas.

Experiments performed under steady-state conditions have given rise to our understanding of the basic features underlying muscle contraction—the sliding filaments, the lengths of which do not change much during contraction, the involvement of cross-bridges between the filaments in the production of the force necessary for sliding, and the use of ATP as the immediate source of chemical energy—but such experiments do not distinguish well between the various theories proposed to account for the steady-state results. The dynamic responses of the muscle to perturbations from the steady state impose much more rigid restrictions on the candidate theories. The last part of this article will discuss such restrictions.

We ought to mention three other recent sources of information on muscle. Aidley (1971) gives an excellent introductory account of muscle. The papers presented at the 1972 Cold Spring Harbor Symposium of Quantitative Biology appear in volume 37 of that series.

Several papers from that volume are cited extensively here. Also, Simmons and Jewell (1973) have recently written an article which is in many respects complementary to this one. In particular they discuss the build-up of activity in muscle following neural stimulation, an aspect with which we do not deal at all.

II. HISTORICAL

1. *Viscoelastic or "Cocked-spring" Theories*

The modern view of muscle contraction originated in the mid 1950s. It was at this time that the sliding-filament theory was first advanced (H. E. Huxley and Hanson, 1954; A. F. Huxley and Niedergerke, 1954; A. F. Huxley, 1957), and experiments showing that the three-element model of A. V. Hill would not account for the non-steady-state mechanics were performed (Jewell and Wilkie, 1958). It was also during this time that the first electron micrographs of muscle were obtained. The purpose of this section is to look at the work that was performed before this time and at the theories current then.

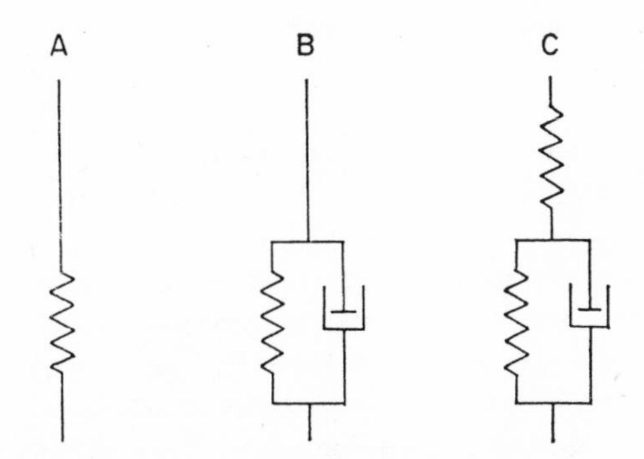

FIG. 1. Viscoelastic or "cocked-spring" models of muscle. A, Simple spring. B, Damped spring. C, Damped spring plus series elasticity.

These earliest formulations stemmed from the suggestion that muscle, when activated, behaved like a stretched spring (Fig. 1 A) (Weber, 1846). This was shown to be too simplified by Blix (1893) who measured the tension exerted by the muscle when it was (a) being stretched, and (b) being released, and showed that, at the same muscle length, the tension exerted by the muscle during stretch was greater than that during release. A simple spring would exert the same tension in these two experiments.

A further suggestion (A. V. Hill, 1922) was that the spring was damped by viscous forces (Fig. 1 B). This would account for Blix's results, and arose from experiments in which the maximum speed of shortening of muscle pulling against different inertial loads was measured for human subjects pulling with their arm muscles. Similar experiments were repeated on isolated frog sartorius muscles by Gasser and Hill (1924), thereby removing the difficulty of interpretation always present in experiments on intact animals caused by the complexity of neurally mediated direct and reflex control. Gasser and Hill also performed quick-release and quick-stretch experiments. Their conclusion was that for all but the fastest stretches their results were consistent with a model composed of a spring damped by viscous forces.

Levin and Wyman (1927), working in Hill's laboratory, demonstrated that a simple viscous spring was insufficient by measuring, with improved apparatus, properties similar to those measured by Gasser and Hill. Levin and Wyman proposed that muscle should be represented by such a damped spring in series with an undamped elasticity (Fig. 1 C). In fact, Gasser and Hill's (1924) results also required such an element since the time course of redevelopment of tension following decreases in length was slow compared to the rate of application of the length change, and for small length changes the tension did not drop to zero as would occur with a damped spring in the absence of any series elasticity.

2. *The Fenn Effect*

All the theories of contraction in muscle mentioned so far are based upon the idea that, following a stimulus, the muscle acts like a cocked spring along with various passive elements (springs and dashpots). A common property of all such models is that the amount of energy available should be independent of the manner of contraction provided that all methods of release of energy are taken into account. These include the performance of mechanical work, the production of heat, and possibly the reconversion of mechanical energy to chemical energy. Fenn (1924) demonstrated that the total energy released in the form of heat plus mechanical work during a single twitch in frog sartorius muscle was greater in twitches during which the muscle was allowed to shorten than during a twitch in which the muscle was held at constant length. This result is known as the Fenn effect; it is of fundamental importance in that it demonstrates that the total release of energy in a twitch can be regulated by mechanical changes in the muscle and is not, as had been thought, an all-or-nothing phenomenon. Strictly speaking, Fenn's results did not in themselves demonstrate this—the conclusion is valid only if no mechanical energy is converted back into chemical energy. If more chemical restoration were to take place during an isometric contraction than during an isotonic contraction, then the total energy released during the contraction could be constant. In fact it seems probable (Curtin and Davies, 1973) that mechanical energy is not reconverted to chemical energy in muscle, and thus that this assumption made in the interpretation of Fenn's results is valid. This problem is usually overlooked, although Levin and Wyman (1927) treat it very clearly.

Assuming that the sum (heat + work) is a good measure of the energy released by muscle, then Fenn's results demonstrate that the cocked-spring models of muscle are inappropriate.

3. *A. V. Hill's Three-element Model*

The next major development in the field was due to a combination of improved mechanical and thermal measurements by A. V. Hill (1938). Attempts were being made in the 1930s to characterize the performance of muscle in terms of the relationship between velocity of shortening and the load on the muscle rather than the total work obtainable from the muscle as a function of the load, which was the relationship aimed at in the 1920s. That is, the emphasis was upon power rather than work. Stevens and Metcalf (1934) suggested that their experiments were consistent with the muscle producing constant mechanical power, i.e. if the load on the muscle is P, giving rise to a velocity of shortening v (which is, for an interval of time, constant for that load), then $Pv = \text{const.}$ Fenn and Marsh (1935), with more accurate measurements, showed that, except at very low speeds, the sum of the external power Pv and an internal power required to overcome a viscous force kv^2 was constant. They showed, however, that the nature of the viscous force was unclear—in particular, two different methods of its measurement gave very different values for its magnitude. Hill's contribution was to measure the experimental relationship between load and velocity and to improve the time resolution of his thermopiles to such an extent that he was able to demonstrate that under a wide variety of mechanical conditions the muscle developed a greater amount of heat whilst shortening than it would have done in the absence of shortening. The magnitude of this heat was dependent only upon the total degree of shortening. From these thermal measurements Hill showed that the heat produced in excess of that during an isometric contraction was proportional to the total shortening x. The constant of proportionality between heat and shortening is called a, so that the rate of heat production during the shortening is $a(dx/dt) = av$, where v is the velocity. The mechanical power is Pv, and thus the rate of release of energy of the muscle allowed to shorten, in excess of that during an isometric contraction, is $(P + a)v$. Experimentally, Hill showed that this rate

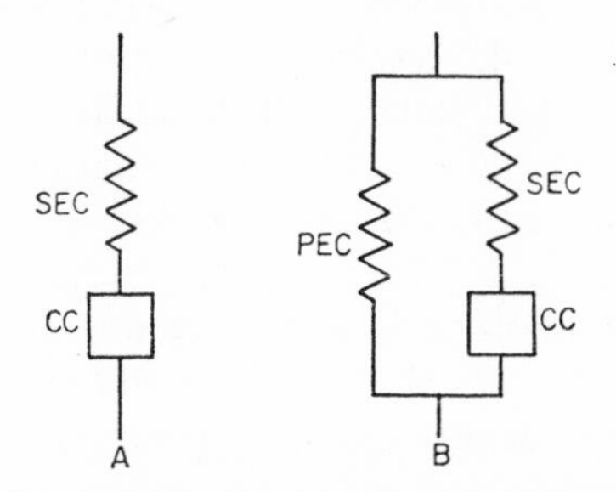

FIG. 2. A. V. Hill's model. A, Two-element model containing a contractile component (CC) in series with a series elastic component (SEC). B, Three-element model, including a parallel elastic component (PEC) to account for the elasticity of the relaxed muscle at considerable stretch.

of extra energy production was related linearly to the load on the muscle, being zero, of course, for the isometric contraction when $P = P_0$, the isometric tension. Accordingly,

$$(P + a)v = b\ (P_0 - P), \tag{1}$$

where b is a new constant of proportionality. The value of b can be found by plotting the excess energy liberated against the load. Of course, the above equation also predicts the relationship between force and velocity, and this can be tested via mechanical experiments alone. The values of the constants so obtained ought to agree with those obtained from the thermal measurements. Hill showed that such agreement was obtained. The equation above, known as the Hill equation for muscle, gives a good description of the behaviour of the constant-velocity properties of muscle.

Hill further suggested in the same paper (1938) that muscle could be adequately represented by an "active" contractile component (CC) whose properties were such that its behaviour was described by eqn. (1), relating force to velocity, in series with an undamped elastic component which became known as the series elastic component (SEC) (Fig. 2 A). The necessity for the series elastic component was inferred from data such as that of Levin and Wyman (1927), showing that there was an instantaneous tension change for applied length changes. Hill's two-element model differed from the viscoelastic models in that the behaviour of the contractile component was active—the amount of energy obtainable from it was dependent upon its mechanical changes, unlike that of the earlier models.

The difference between active muscle and relaxed muscle in this system was that in relaxed muscle the contractile component contributed negligible force. Activity was caused by a build-up of the active properties of the contractile component very rapidly after stimulation. Hence there ought not to be tension in relaxed muscle. Since large extensions (to lengths greater than about 1.4 times the resting length in the body) do result in significant tension in the relaxed muscle, a third component in parallel with the other two, the parallel elastic component (PEC), was added (Fig. 2 B). The relationship between length and tension for this element was simply that found experimentally for relaxed muscle.

Note that in the region of normal body length the PEC does not contribute to the tension of the muscle, and can be ignored. The model is often referred to as a two-element model under these conditions.

The series elastic element is characterized by a length–tension diagram. The relationship can be determined, given the PEC, experimentally either by applying quick stretches and releases of varying magnitudes and measuring the tension changes, or by applying sudden changes of tension and measuring the resulting instantaneous length changes.

This three-element model of Hill dominated the ideas of muscle for many years. It is capable of describing many of the results obtainable from vertebrate striated muscle, both mechanical and thermal, fairly accurately and it is easy to understand. For example, the model, with parameters determined from independent measurements, will predict qualitatively the rate of rise of tension in an isometric tetanus. It is known, from the fact that quick stretches applied very soon after the stimulus result in the full isometric tension, that muscle is active (in the sense of being capable of contraction) a long time (tens of milliseconds) before the tension reaches its maximum value. This is accounted for in Hill's formulation by the fact that before the contraction starts the tension is zero, and the SEC is therefore at its rest length. As tension develops, the length of the SEC increases, allowing a shortening of the CC to take place. Thus, in the terms of the model, the time taken for the development of the full isometric tension was the time required for the CC to stretch the SEC to the length appropriate for this tension. This time is considerable and within the resolution of the experimental results at that time.

There are several other aspects of Hill's three-element model worth mentioning. At the time it was proposed it was a major breakthrough in that it tolled the death of the visco-elastic theories. Fenn in particular had appreciated this, and Hill's (1938) paper had the necessary experimental improvements and his very lucid explanations. His theory, however, has important drawbacks: it neither treats the way in which the two series elements of the model are related structurally, nor does it invoke any candidate mechanism underlying the properties of the CC. The formula, despite the manner of its development, is empirical and has not helped specifically toward understanding of the CC. For this reason its usefulness is limited. Of course, any theory that is developed must relate force and velocity in a similar way, and it is useful to have a convenient mathematical expression to describe the relationship, as Hill's equation certainly does.

4. *A Critical Test of A. V. Hill's Three-element Model*

About 20 years later, improvement in experimental technique enabled a critical test of Hill's model. Jewell and Wilkie (1958) measured the characteristic properties of the CC and the SEC, below that muscle length at which the PEC contributes tension, and on the same muscles measured both the initial development of tension in an isometric tetanus, and the redevelopment of tension during tetanus following very rapid release by just that amount necessary to cause the tension in the muscle to fall to zero. They then predicted the time-course of the development of tension on the two-element model using the measured properties of the CC and the SEC.

The properties of the two components were measured in isotonic-release experiments. In such an experiment the muscle is tetanized and allowed to develop its full isometric tension. The muscle is held isometric by means of a stop (Fig. 3). When the stop is removed the

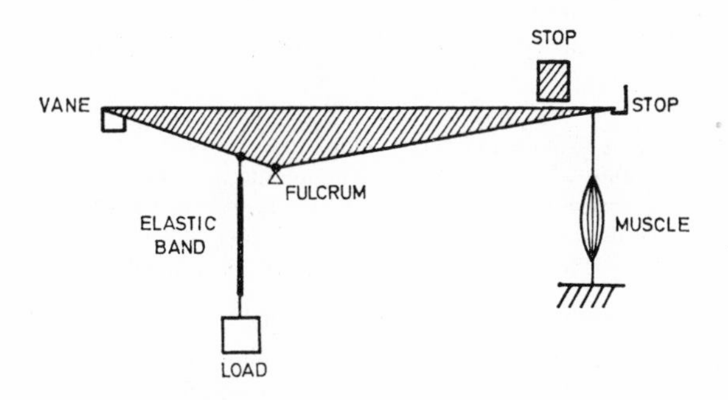

FIG. 3. Schematic diagram of the Jewell and Wilkie apparatus. The muscle is attached to the arm of a balance which can be loaded by means of weights attached to the balance via an elastic band (to reduce effects due to acceleration of the mass). Length changes in the muscle are monitored by means of a vane which partially obscures a light beam falling on a photocell. The stop above the arm keeps the mass from extending the relaxed muscle; removal of the stop at the end of the arm releases the muscle at a chosen time during a tetanus, resulting in the tension–step experiment of Fig. 4 A.

muscle is then loaded only by the inertia of the balance arm, which is very small, and the weight suspended from this by means of a long, compliant, elastic band. Inertial forces were minimized (note the use of the elastic link in Fig. 3) so that the muscle encountered a virtually constant force from the moment of release of the stop which was holding it isometric. The general form of the tension and length records is illustrated in Fig. 4 A, and a typical recording of the length changes from Jewell and Wilkie's paper is shown in Fig. 5. Notice that there is an initial sudden length change which, as far as can be determined, occurs at the same time as the tension change. This is what would be expected from an elasticity, and this part of the curve, on the Hill view, was due to the SEC. After this elastic change the length record settles down, after a few oscillations, to a constant velocity. This is the velocity relevant to the particular force loading the muscle, and gives one point on the characteristic curve of the CC. Exactly what is happening in the period when the length record is oscillating is not clear because these are, in the main at least, spurious oscillations caused by imperfections in the apparatus. However, if the two-element model is correct the actual response would be that shown by the line *xyz*, which is an extrapolation of the constant velocity back to a predicted elastic change of length. Jewell and Wilkie performed a series of such experiments with different loads, giving the series of curves shown in Fig. 6. From these curves the properties of the hypothetical SEC and the CC can be determined. For each record, by extrapolating the length record as before, values for both the elastic length change (*Oz* in Fig. 5) and the velocity (the slope of *zyx*) can be measured relevant to the

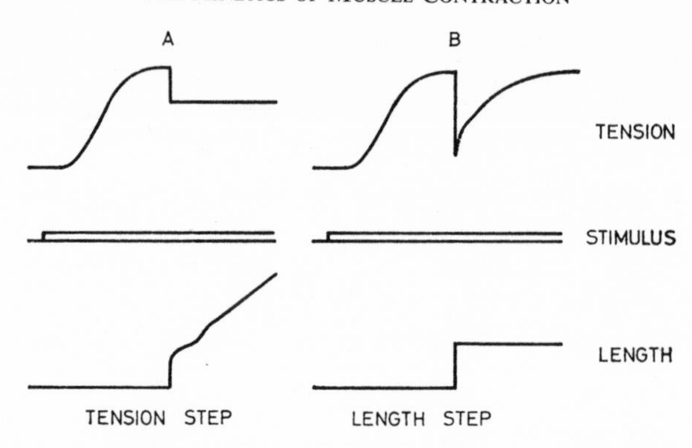

FIG. 4. Illustrations of the two main types of mechanical experiment performed on active muscle during tetanus. In both cases the full tetanic tension is first developed under isometric (constant length) conditions. In tension–step experiments the tension is then changed rapidly to a new level, and the resulting length changes of the muscle recorded. In length–step experiments the length of the muscle is forced to a new value very rapidly, and the resulting tension changes followed. Decrease of length, upward.

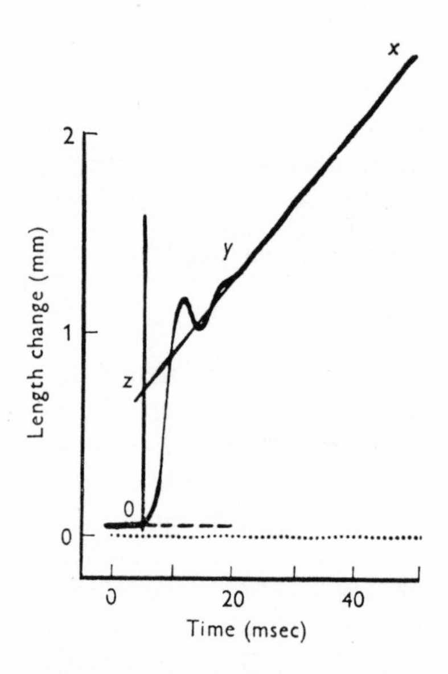

FIG. 5. Length change resulting from a step reduction of tension, as described in the text. Decrease of length, upward. (From Jewell and Wilkie, 1958.)

particular force (or tension) applied to the muscle. Each curve gives one point on the force–velocity curve of the CC and one point on the tension–length curve of the SEC.

Figure 4 B shows in diagrammatic form the way in which the isometric redevelopment of tension was measured. The time-course for the redevelopment of tension following a release can now be predicted from the curves giving the properties of the SEC and the CC, assuming that the constant-velocity CC characteristic applies immediately after the step. Figure 7 shows Jewell and Wilkie's figure comparing the predicted time-course with that measured experimentally under two conditions: (a) during the initial development of tension after

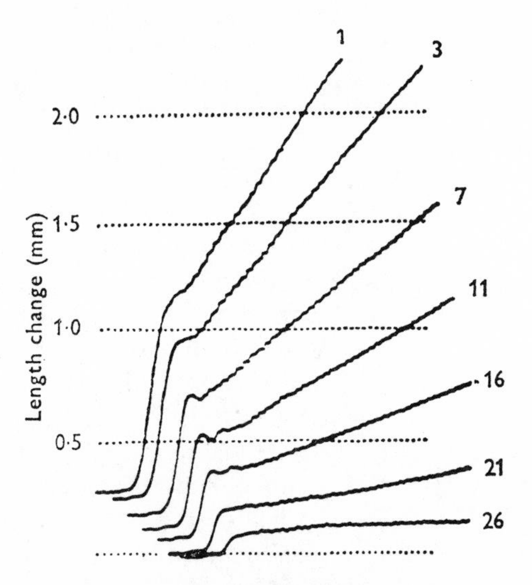

FIG. 6. Length changes (decreases of length upwards) resulting from a series of tension–step experiments (as in Fig. 4 A) in which the tension was lowered suddenly from the full isometric tension to the tension (values in gram-wt.) shown alongside the trace. The origins of the traces have been shifted for clarity. The dots on the calibration lines denote time intervals of 1 msec. (From Jewell and Wilkie, 1958.)

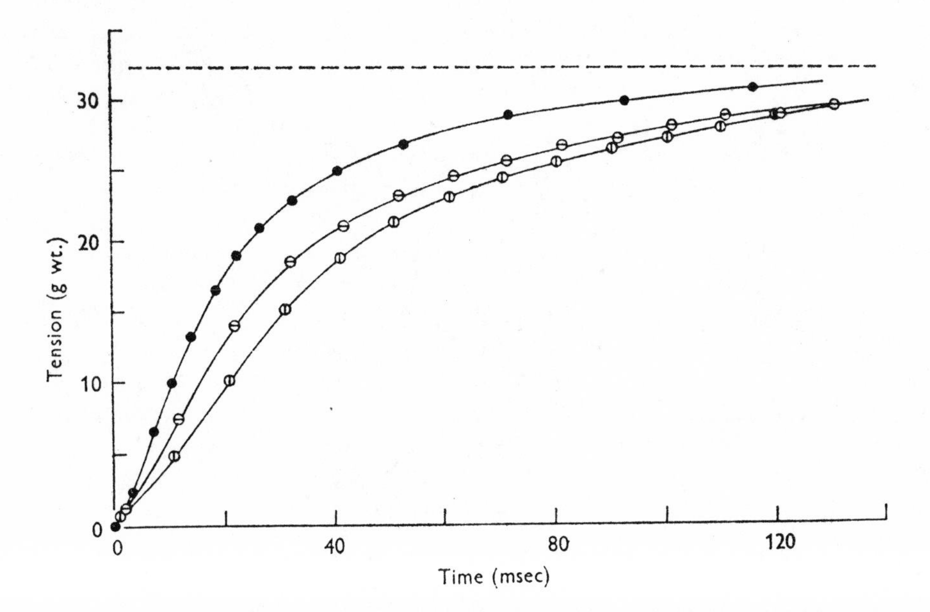

FIG. 7. The open circles denote the measured development of tension, ⦶ during the rising phase of a tetanus, and ⊖ following a quick release, during a tetanus, just sufficient to cause the tension to fall to zero. The closed circles show the predicted time course of development of tension on Hill's two-element model, as discussed in the text. (From Jewell and Wilkie, 1958.)

stimulation, and (b) after application of the sudden length change. The errors in the measurements and in the predictions are less than the size of the points. The difference between prediction and results is considerable (e.g. after 20 msec the predicted tension is 55% of the final tension, whereas the measured tension is only 40%). The two-element model, as proposed by Hill, cannot account for the results of this test. Something fairly major is wrong with it.

The failure could lie in the possibility that the early response in Fig. 5, masked by an artifact and predicted via Hill's model to be the extrapolated straight lines, is actually very much different. In fact, any formulation which assumes that the constant-velocity force-velocity characteristic is attained instantaneously following the step, and that the properties of the SEC are invariant, will produce similar discrepancies. As we shall see below, models for which this assumption does not arise (e.g. A. F. Huxley's (1957) model as treated by Julian, 1969) readily account for Jewell and Wilkie's data. In another sense, the failure of Jewell and Wilkie to account for their results with Hill's model using the above assumption focuses attention upon the need to measure the very early events in such mechanical transients more accurately, progress in which will also be described below.

Pringle (1960) suggested that the degree of "activation" in muscle might be affected by the muscle tension, and that the discrepancy between the Hill two-element model and the results of Jewell and Wilkie (1958) might be explained by this means. Although this has not been pursued we shall meet the idea of such activation again in § VII.

III. SOME STRUCTURAL CONSIDERATIONS. THE SLIDING-FILAMENT THEORY

Since Hill's model was formulated in the late 1930s it is obviously not based upon the structure of muscle which was being discovered in the 1950s, and which forms the basis of our present knowledge. We shall outline the salient features of muscle structure here, despite the amount that has been written on this subject. An excellent general account of the structure is given by H. E. Huxley (1971).

The main change in understanding arose from structural studies made of the repeating pattern (sarcomere) under the light microscope using polarized light, phase contrast, and interference techniques by H. E. Huxley and Hanson (1954) and A. F. Huxley and Niedergerke (1954). The technical capabilities for the required observations had existed for many years, and it is perhaps surprising that they were not made earlier. The observations, now well known, that when the length of the muscle fibres was changed the striation pattern changed in such a way that two of the lengths within the sarcomere remained constant, resulted in the deduction that muscle worked by the relative sliding of two sets of interdigitating filaments. This was soon shown to be correct by electron microscopy (H. E. Huxley, 1953). The implication from the light-microscope observations was that the filament lengths did not change during contraction, and this was confirmed to a greater degree of resolution some years later by electron microscopy (Page and H. E. Huxley, 1963).

This finding, known as the sliding-filament theory, was a major development over the previous view (e.g. Polissar, 1952; Morales, 1959) that shortening in muscle occurred by a folding up of long-chain molecules, which formed a network from one end of the muscle to the other.

The central problem in understanding contraction now became capable of being formulated far more precisely—What makes the filaments slide past one another and how does it work?

Very soon after the sliding filaments were proposed, A. F. Huxley (1957) suggested that some kind of mechanical link which could "make and break" might be responsible for the production of the force between the filaments required for sliding; H. E. Huxley (1957), on the evidence of the side projections from the filaments seen in the electron micrographs, also suggested that these might be mechanical linkages producing such a force. Other ideas (see § VI) for the way in which the sliding force might be produced have not involved such mechanical links.

The electron micrographs showed that the filaments were of two types—thick (A) and thin (I) filaments. By various extraction procedures Hanson and H. E. Huxley showed that the A filaments contained myosin and the I filaments actin (Hanson and H. E. Huxley,

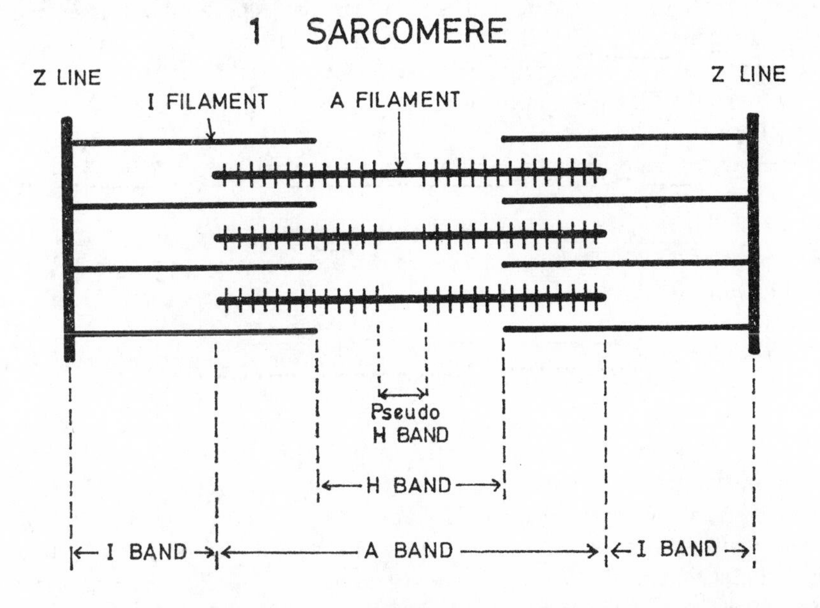

FIG. 8. Arrangement of filaments within a sarcomere, defining the terminology used. The sarcomere is delimited by the *Z* lines (sometimes known as *Z* discs). Note that the basic unit of contraction is the half-sarcomere, between the *Z* line and the centre of the sarcomere (often identifiable by the presence of extra material known as the *M* line). Note also the distinction between the *H* zone (the region not occupied by I filaments) and the pseudo-*H* zone (corresponding to the region of A filament not containing cross bridges (the clear zone).

1957; H. E. Huxley and Hanson, 1957). The result was confirmed by reprecipitating the dissolved proteins and showing that the structures thus formed were similar to the respective filaments (Hanson and Lowy, 1963; H. E. Huxley, 1963). More recently it has been shown that there are other proteins associated with the two filaments. The I filaments contain tropomyosin and troponin (Pepe, 1966; Caspar *et al.*, 1969; Moore *et al.*, 1970), and the A filaments contain C protein (Offer, 1973; Pepe, 1973). The filaments are arranged into a sarcomere (Fig. 8) delimited by the *Z* disc, which probably contains the protein α-actinin (Stromer and Goll, 1972).

Myosin is a long match-shaped molecule, easily split by light tryptic digestion into two parts: (1) a rod about 90 nm long, formed of a two-chain α-helix, known as light meromyosin (LMM) and of molecular weight about 150,000 Daltons; LMM can aggregate into filaments resembling the backbone of the A filament. (2) a heavier part (molecular weight

350,000) known as heavy meromyosin (HMM) consisting of two globular "heads" (about 10 nm long) with a tail (50 nm long) known as subfragment 2 (S_2). Most of the "tail" can be separated from the globular "heads" known as subfragment 1 (S_1). Each myosin molecule contains two S_1 heads and one S_2 tail.

The thick filaments, containing the myosin, have side projections, known as cross-bridges, at regular intervals along their length, except for a "clear" projection-free region about 0.2 μm long at their centre. Myosin, dissolved from muscle in solutions of high ionic strength and then reprecipitated, forms similar structures of varying lengths, but always containing the clear region, suggesting that the filament is composed of an aggregation of molecules which are polarized in opposite directions on either side of the central point. LMM aggregates to form filaments, but without the projections, thus suggesting that the cross-bridges are composed of HMM. The solubility properties of S_2 have led Lowey *et al.* (1966) to suggest that this part of the molecule is not a part of the backbone but provides a "hinge" between the filament and the globular heads, enabling these heads to swing away from the backbone.

The ATPase activity and the actin binding sites are located on the S_1 heads of the myosin molecule (Young, 1967).

Actin is a globular protein, approximately spherical with a radius of about 5.5 nm and a molecular weight of 45,000 Daltons. A large part of its primary sequence has been determined (Elzinga and Collins, 1973). It aggregates to form filaments, most easily described as two chains of monomers twisted round one another to form a twin-threaded coil with a complete twist every 13–15 monomers. Tropomyosin is a rod-shaped molecule 41 nm long and with a molecular weight about 63,000 Daltons (Cohen *et al.*, 1971). It is a double α-helix. By analysing the optical diffraction of electron micrographs of I filaments, Moore *et al.* (1970) have shown that the tropomyosin probably lies along the length of the I filament, within the groove between the two chains of actin monomers. X-ray diffraction studies (H. E. Huxley, 1973) have suggested that the activation of muscle via calcium results in a movement of the tropomyosin around the actin monomers, mediated by the troponin molecules (see § IV), the implication being that myosin binding sites on the actin are thereby uncovered, enabling actin to activate the myosin ATPase, which is otherwise prevented.

It is important to keep in mind the geometry involved. Figure 9 A illustrates the general shapes of the various molecules drawn to scale, with various dimensions indicated. Figure 9 B shows a small section of the complete structure, also drawn to scale, to illustrate the probable way the proteins fit together. The filament separation d in intact vertebrate muscle increases as the sarcomere length l decreases in such a way that the sarcomere volume remains constant, i.e. $ld^2 = \text{const.}$ (Elliott *et al.*, 1963). The diagram has been drawn for two different sarcomere lengths—one for the situation where there is very little overlap, and the other for complete overlap of the A filaments. Notice that if the S_2 part of the myosin molecule can act as suggested earlier, then the large change in spacing between the surfaces of the A and I filaments can be easily accommodated, as suggested by H. E. Huxley (1969).

The arrangement of the filaments and the relevant dimensions within the sarcomere are shown in Fig. 10 for both vertebrate striated muscle and the flight muscle of *Lethocerus cordofanus*. The diagrams are drawn for approximately normal body length. Notice the different arrangement of the I filaments around the A filaments in the two muscles and the very short I-band region in the insect muscle.

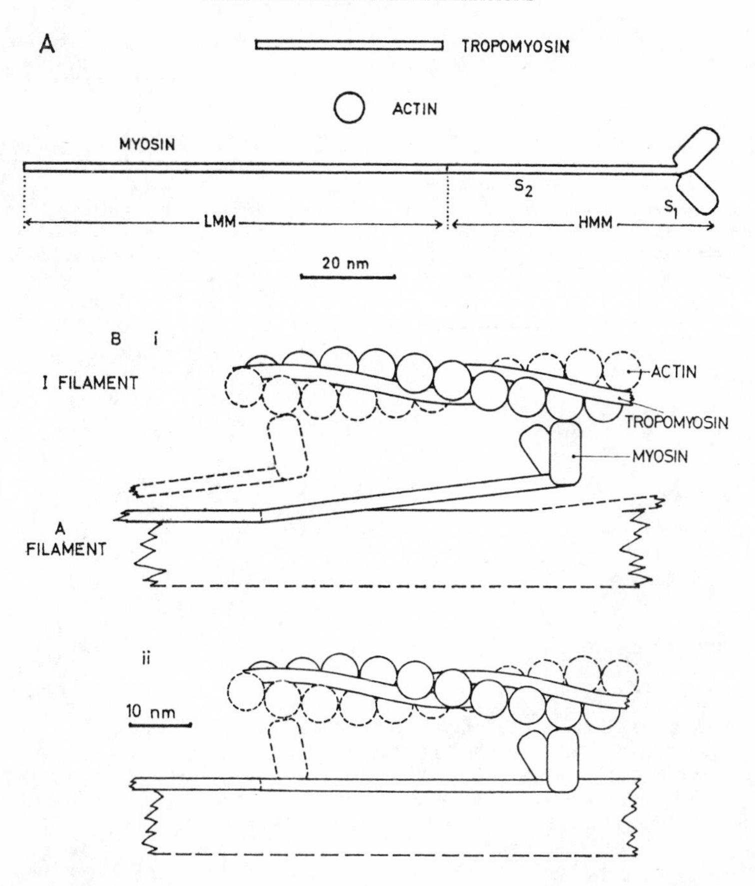

FIG. 9. A, Actin, myosin, and tropomyosin molecules, drawn to scale. B, Arrangement of the actin, tropomyosin, and myosin molecules in the A and I filaments. Note that the troponin (which is bound to the tropomyosin) is not included in the diagram. Two different interfilament spacings are drawn, corresponding (in frog muscle) to sarcomere lengths of (i) 2.0 μm (A–I filament spacing, centre-to-centre, 26 nm), and (ii) 3.6 μm (spacing, 19 nm) (Elliott *et al.*, 1963).

One further structure within the sarcomere of the insect muscle has been suggested on several occasions (Auber and Couteaux, 1963; Garamvolgyi, 1969; Zebe *et al.*, 1968; Reedy, 1972). This concerns a connection between the end of the A filament and the Z disc, often named a connecting or C filament. Such a structure could account for the high stiffness of relaxed insect fibrillar flight muscle as compared with vertebrate muscle. Ashhurst (1971) gives an excellent critical account suggesting that in fact no clear evidence for the existence of such a connection has yet been published. We agree with her discussion. However, if the insect muscle is put into rigor and then stretched by about 10%, the I filaments break away from the *Z* line in some sarcomeres leaving a "clear" zone between the *Z* line and the interlocked A and I filaments. There is still structural continuity between the *Z* line and the A–I filament array. This can be inferred from the fact that many *Z* discs have clear zones on both sides. This could not occur in the absence of continuity, because otherwise whichever side broke first would be unable to exert any force to pull the *Z* line from the adjacent

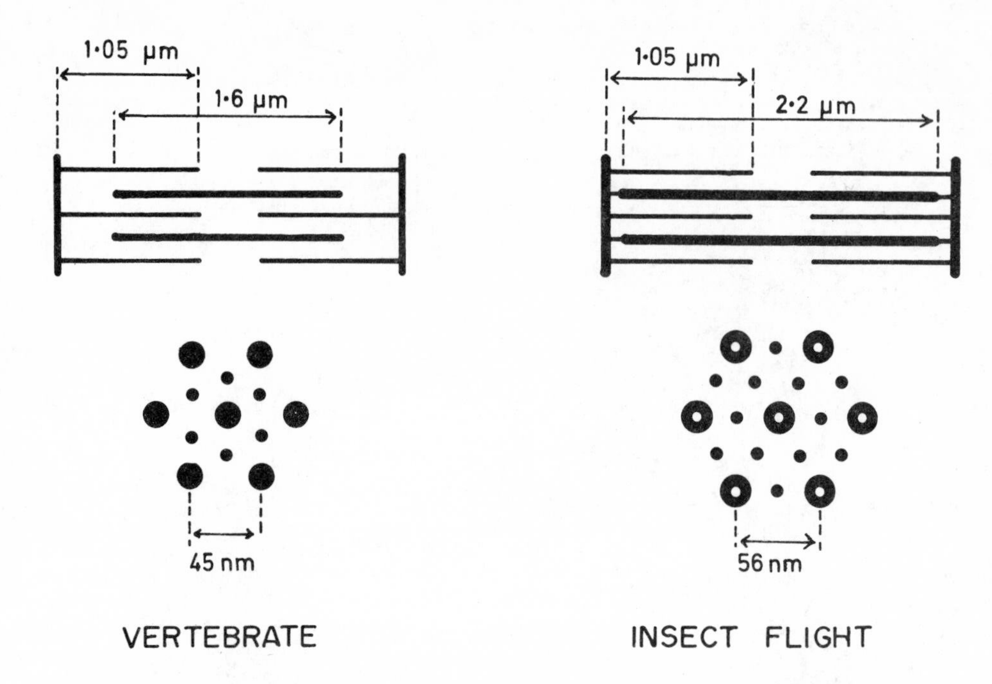

FIG. 10. Comparison of the structure of vertebrate striated and insect fibrillar flight muscles. Both are striated with thick (A) and thin (I) filaments forming sarcomeres. At body length the A filaments of vertebrate muscle occupy about 66% of the length of the sarcomere, whereas the insect fibrillar flight muscle has A filaments extending almost from one *Z* line to the next. As shown in Figs. 11 and 12, there appears to be structural continuity between the A filament and the *Z* line in the latter. The A filaments of the insect muscle are thicker than those of the vertebrate muscle, and contain a central core which does not stain in sections prepared for electron microscopy. The arrangement of I filaments around the A filaments is different in the two muscles, the ratio of I to A filaments being 2:1 in vertebrate and 3:1 in insect flight muscle.

array of A and I filaments. Furthermore, there is evidence in the electron micrographs of such a structure. Figures 11 and 12 show longitudinal and transverse sections, respectively, through such a preparation. There are connections between the *Z* line and the A filaments. That the connections are to the A filaments can be seen most easily in the transverse sections by viewing them obliquely. The alignment is very striking. Reedy (1972) shows a longitudinal section of such a preparation, but this by itself is not as convincing as Figures 11 and 12. We cannot at the moment exclude the possibility that these strands may comprise actin or *Z*-line material that has been pulled away by the A filaments under these conditions.

One point of essential terminology is not used consistently in the literature. This concerns the number of cross-bridges. In both vertebrate striated muscle and insect flight muscle the axial repeat of the A filament is about 14.5 nm. For vertebrate muscle, H. E. Huxley and Brown (1967) suggest that two (X-ray and e.m. observed) "cross-bridges" arise every 14.5 nm from opposite sides of the filament. But calculations of the myosin content of the A filament suggest that each of these "cross-bridges" ought to be composed of two myosin molecules, i.e., four S_1's (see, Squire, 1971). Similar estimates for the insect muscle indicate that three myosin molecules arise from each cross-bridge site (Chaplain and Tregear, 1966). Ought one to call each myosin molecule a cross-bridge, and thus have multiple cross-bridges

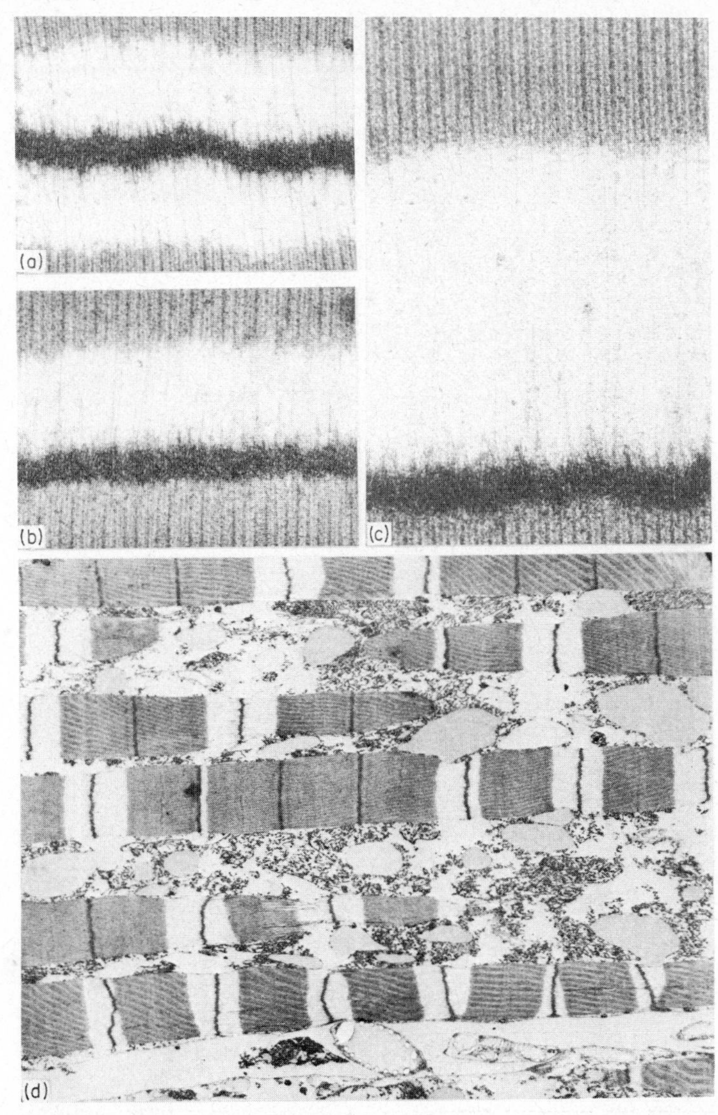

FIG. 11. Glycerol-extracted dorsal longitudinal flight muscles from *Lethocerus cordofanus*, fixed in gluteraldehyde, postfixed in osmium, and stained with lead and potassium permanganate as described by Reedy (1967). The muscles were stretched from rest length whilst in rigor (see White, 1969) by about 10%. The stretch takes the fibres beyond their elastic limit under these conditions. The myofibrils break at the junction of the I filaments and the *Z* line, leaving "clear" zones. These clear zones are crossed by electron-dense strands which seem to originate from the A filaments (but in these thick sections it cannot be excluded that the structures do not originate from I filaments above or below thick filaments). These structures have been called "C filaments" (White, 1967). Note that the clear zones are almost never seen with widths greater than about 1 μm, and that along one myofibril there may be many clear zones. This result would not be expected unless there were some mechanical continuity between the *Z* line and the A–I filament matrix. (a and b: ×40,000. c: ×50,000. d: ×6500.)

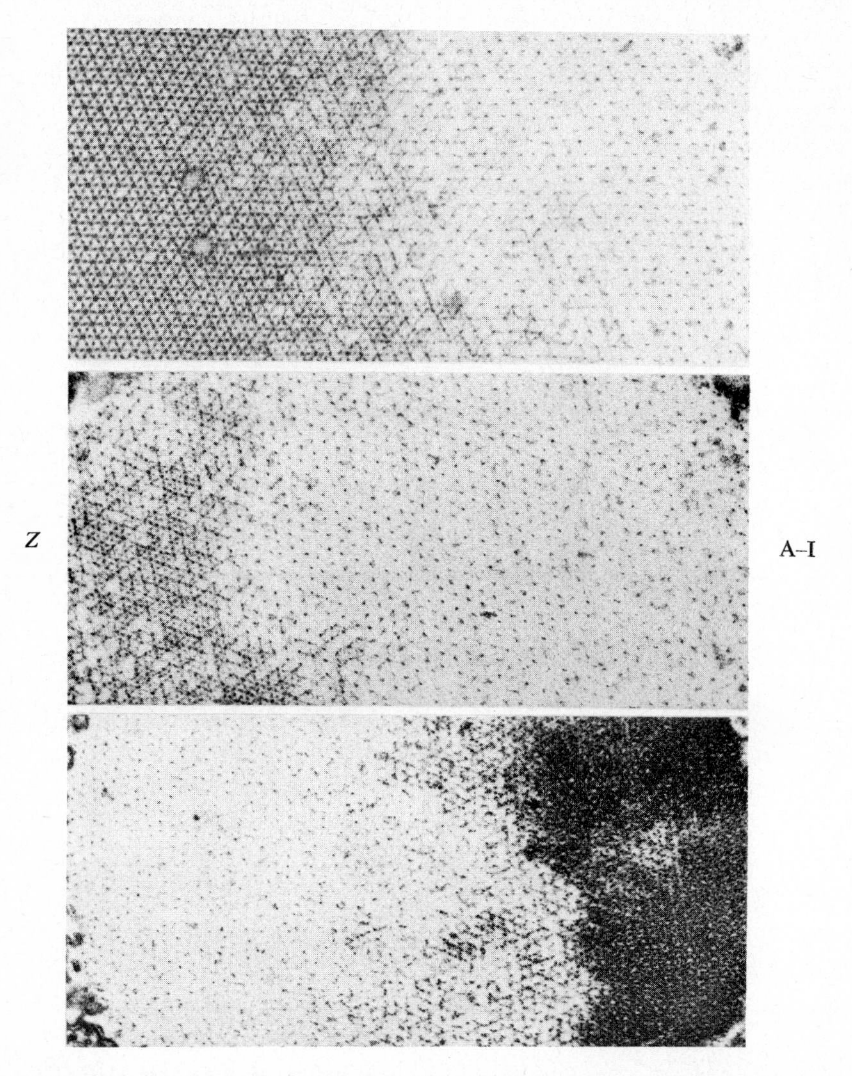

FIG. 12. Transverse sections of the preparation of Fig. 11. The sections are slightly oblique, and include the clear zone. The strands of Fig. 11 can be seen clearly. Close to the A–I filament matrix their arrangement is hexagonal, and by viewing the micrographs obliquely one can see that they are aligned with the A and not the I filaments. Near the *Z* line their arrangement becomes disorganized. It is suggested that these structures may form a mechanical continuity between the *Z* lines and the A–I filament matrix; such a connection could account for the high resting elasticity of insect fibrillar flight muscle (White, 1967). (×60,000.)

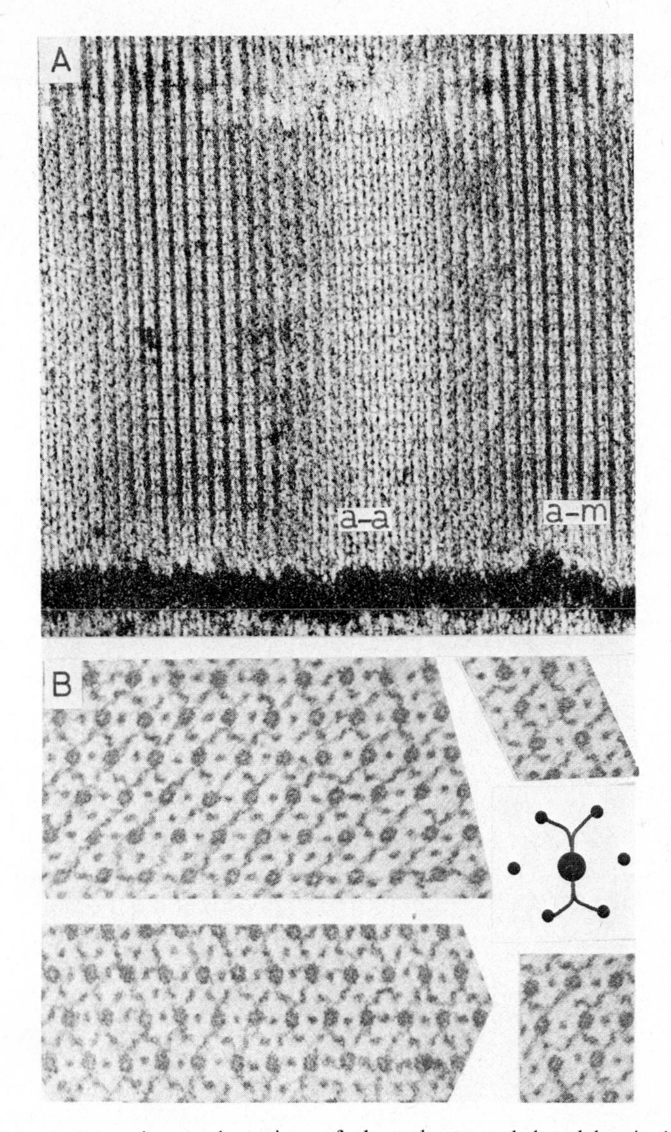

FIG. 13. Thin electron-micrograph sections of glycerol-extracted dorsal longitudinal flight muscle from *Lethocerus*, prepared by Dr. M. K. Reedy. A, Longitudinal section, sufficiently thin that only a single layer of filaments is seen: in different regions the section cuts through a layer containing I filaments only (*a-a*) and a layer containing A and I filaments (*a-m*). Cross-bridges, angled at about 45° to the filament axis, can be seen clearly. The section includes the *Z* line and the *M* line of one half-sarcomere. (×82,000.) B, Thin transverse sections from the same material, showing the "flared-*X*" form of the cross-bridges. The sections are sufficiently thin that the arms of the *X* all originate from the two cross-bridge origination sites on opposite sides of the A filament. (×205,000.) (From Reedy, 1968.)

arising at one "site", or perhaps each "site" a cross-bridge and thus have several molecules per cross-bridge? Here, we have for the moment adopted the former, and thus say that in vertebrate muscle there are two cross-bridge origination sites on opposite sides of the A filament every 14.5 nm, each site having two cross-bridges, and that in insect flight muscle there are also two origination sites every 14.5 nm, but that there are three cross-bridges arising from each site. In the insect flight muscle, Reedy (1968) showed that two of the three cross-bridges from one site could attach to different I filaments (Fig. 13).

With this knowledge of the structure, we are in a position to evaluate the present evidence that cross-bridges are the elements producing interfilament shear force.

(1) Ramsey and Street (1940) showed that the active tension which could be developed in isometric contractions by muscle was strongly dependent upon the length of the muscle, being maximal, for frog fibres, at about the rest length of the muscle in the body, and decreasing on either side to become zero, very approximately, at one half and at just less than double this length. Gordon *et al.* (1966) repeated this experiment with single fibres attached to an apparatus which used markers on the central part of the fibre to control the length of just this central region. By doing this they removed effects both of non-uniformity in the muscle fibre, in sarcomeres near the end of the fibre, and of elasticity in the attachment of the fibres to the apparatus and in the apparatus itself. A summary of the results is shown in Fig. 14, and below this is drawn the appearance of the sarcomeres at particular sarcomere lengths. Notice that the tension development becomes virtually zero when the A and I filaments reach the point of no overlap, that between this point and the sarcomere length at which all cross-bridges on the A filament are overlapped by I filament there is a linear increase in isometric tension, and that when the sarcomere becomes slightly shorter than this, such that there is no change in the number of cross-bridges overlapped, although the sarcomere length is still decreasing, the tension that can be developed by the fibres remains constant. At still shorter sarcomere lengths the I filaments from one side start to interfere with those from the other, and to enter the region of cross-bridges on the other side; the isometric tension falls. This experiment is accounted for in detail by the hypothesis that the cross-bridges are the structural elements of contraction in that the degree of overlap of the I filaments with the region of the A filament from which cross-bridges originate (i.e. not the central clear zone) correlates with tension.

(2) The primary source of chemical energy for the contraction of muscle is adenosine triphosphate (ATP) (Davies, 1964; Kushmerick *et al.*, 1969). Although actin can be made to act as an ATPase under severe treatment such as ultrasonification (Asakura *et al.*, 1963; Barany and Finkelman, 1962, 1963), the turnover of the ADP bound to F-actin is normally very low, even in active muscle (Szent-Gyorgyi and Prior, 1966); the significant hydrolysis of ATP occurs at the myosin site, which is on the S_1 portion of the molecule, i.e. on the globular region of the cross-bridge. It is known that in rigor (in the absence of ATP) the cross-bridge can bind to the actin. This can be seen in electron micrographs of intact muscle fixed in rigor (Reedy *et al.*, 1965) and in the preparation of isolated I filaments "decorated" either with HMM or S_1 (see, for example, Moore *et al.*, 1970). Furthermore, actomyosin superprecipitates under rigor conditions (Maruyama and Gergely, 1962). It is generally maintained, but not proven, that the superprecipitation of actomyosin in solution is the correlate of contraction in intact muscle. Intact muscle contracts as it goes into rigor (White, 1970).

It seems certain that the cross-bridge is involved in the transfer of chemical to mechanical energy because it is the site of hydrolysis of the ATP. It does not necessarily follow that the mechanical force in active muscle is transmitted through an attachment of the cross-

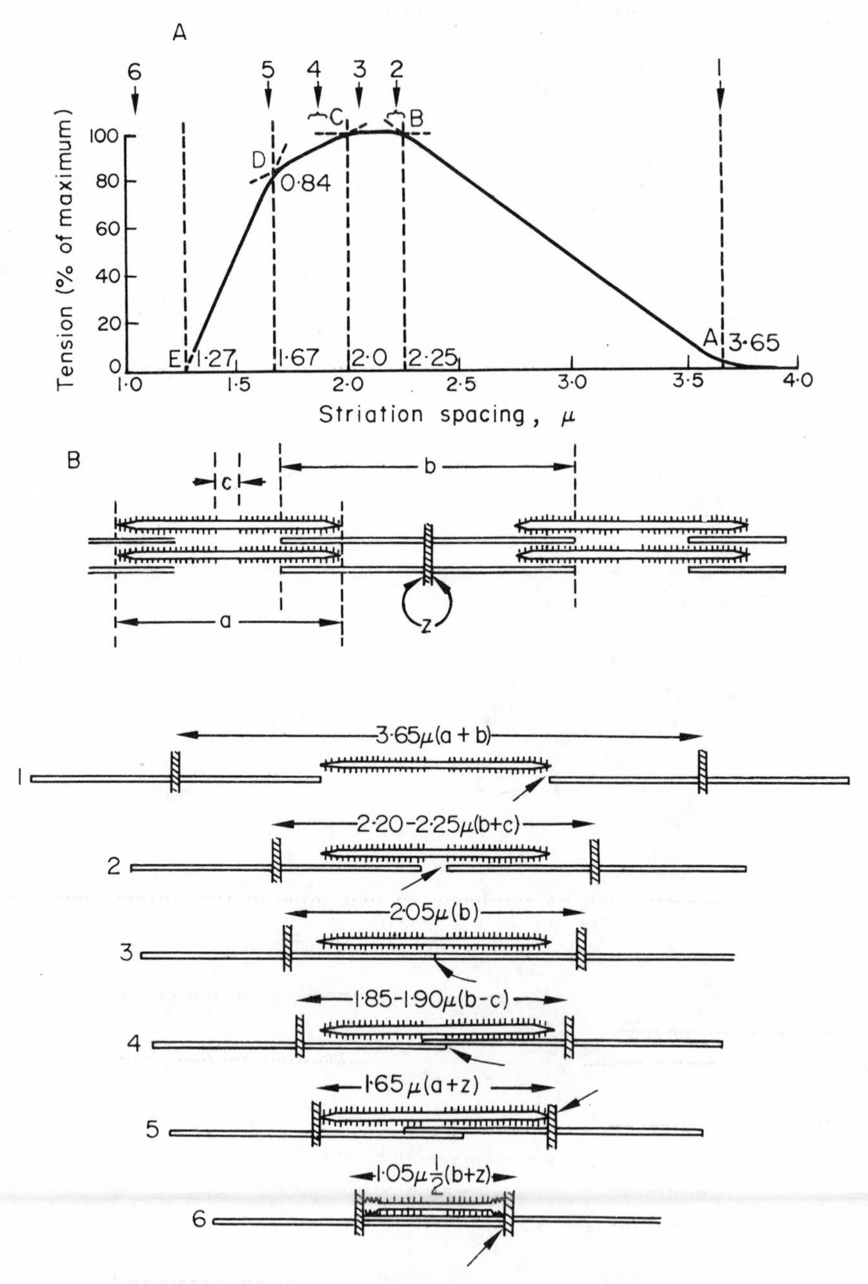

FIG. 14. A. Isometric tension as a function of sarcomere length, in frog semitendinosus muscle. (From Gordon *et al.*, 1966.) B. Appearance of the sarcomere at the six sarcomere lengths marked in part A to illustrate the dependence of tension upon overlap of the thin filaments in the region of crossbridges on the A filament.

bridge, even though it is known that the cross-bridge can bind to the actin in the absence of ATP. However, H. E. Huxley and Brown (1967) suggest from X-ray evidence that during isometric contraction in frog muscle about 20% of the cross-bridges are attached at any instant of time, and Armitage *et al.* (1973) suggest a 10–30% attachment of cross-bridges in *Lethocerus* during development of high tension (see § VII).

(3) There are several techniques for investigating structural changes occurring at the molecular level in intact muscle. The most widespread is X-ray diffraction, but light scattering (Carlson *et al.*, 1973) and fluorescence depolarization (dos Remedios *et al.*, 1972) have also been used. All these show that the cross-bridges depart somehow during contraction from their configuration in relaxed muscle. In frog muscle the situation is perhaps best described by H. E. Huxley (1971): The X-ray experiments on actively contracting muscle "show that upon contraction the cross-bridges within any given thick filament move from the highly ordered helical arrangements present in a relaxed muscle to a much more disordered arrangement in contracting muscle. The experiments show that there is some longitudinal displacement or tilting movement of the bridges and that a considerable amount of radial and/or circumferential movement takes place. This rules out any contraction mechanism in which the bridges remain stationary throughout contraction. However what the experiment does *not* prove is that the bridges move only when they are attached to actin; and there are indications now from very recent experiments by Hazlegrove and myself that activation may itself cause bridge movement, even in the non-overlap zone." This last point is discussed further by H. E. Huxley (1973). We would add here only that activated movement by unattached bridges in the non-overlap zone, while of the greatest significance, is, of course, not necessarily commensurate with shear-force production under the mechanical constraints of attachment.

IV. BIOCHEMICAL

1. *Preliminary Comments*

Biochemical analysis of intact muscle has encountered two difficulties. First, even with single fibres, diffusion can be insufficient to flatten concentration gradients caused by hydrolysis and release of various species. Second, there is no way in which a single protein of the contractile apparatus can be extracted, so that none of that protein remains and no other protein is extracted. Hence most quantitative experiments on the contractile proteins have been done in solution, and unfortunately, do not incorporate the mechanical influences which are perhaps the very essence of the operation of muscle. This lack means that parallel studies of the mechanical and biochemical dynamics, together with the dialogue between them now being developed in terms of conceptual models, are a particularly exciting area of investigation.

Nearly all the biochemical results have been obtained from the rabbit (one rabbit provides enough muscle for protein extractions whereas one frog does not), and nearly all the mechanical work on vertebrate striated muscle has used the frog. The extent to which comparisons can be made is not known, although it is likely that the qualitative conclusions are valid.

A characteristic property of the contractile proteins in solution is their ability to undergo synaeresis or superprecipitation. Of the proteins present, just two—actin and myosin—are required for this phenomenon. If extracted and combined in the absence of ATP then, at physiological ionic strengths (about 0.1 M), actin and myosin combine to form an actin–myosin gel, which is a partially random network of actin and myosin filaments. The two sets of filaments are locked rigidly together because, in the absence of ATP, the myosin molecule can bind to actin in what is termed a rigor link (this link is formed in rigor mortis in connection with the depletion of ATP). If now the ionic strength is raised (above about 0.6 M), myosin dissolves, i.e. the filaments dissociate, and provided ATP is present the gel clears. If ATP is added to the actomyosin gel at low ionic strength the gel first clears, and later superprecipitates. Superprecipitation occurs only when the ATP concentration in the gel is reduced to about 0.3 mM (Eisenberg and Moos, 1967). Thus the duration of the clear phase before the onset of superprecipitation will depend upon the ATPase activity during the clear phase, and can be used as an approximate measure of activity. Table 1, suggested by Evan Eisenberg, summarizes the relationships in more detail.

TABLE 1. ACTIN–MYOSIN INTERACTIONS

	ATP present	No ATP	Low ATP
High KCl 0.6 M	Myosin monomers Actin filaments Completely dissociated	Myosin monomers *bound* to actin filaments (arrow-heads)	Same as ATP present
Low KCl 0.2 M	Myosin filaments and actin filaments dissociated during "clearing"	Actomyosin gel Myosin may be short filaments	Actin and myosin filaments interacting during super-precipitation

The ATPase activity of the myosin during the process of superprecipitation is very high. In a system which superprecipitates, the conditions of the experiment are changing very rapidly, making a study of steady-state effects impossible. Because of this difficulty, earlier experiments investigating the ATPase activity of myosin were performed at high ionic strength (greater than 0.7 M) under which conditions the actin is present in the form of filaments, but the myosin is in solution. These experiments measure the properties of the myosin ATPase, since the actin and myosin are completely dissociated by ATP at these ionic strengths.

2. *The Control of Activity*

The measure of ATPase activity given by the duration of the clear phase has clarified the roles of tropomyosin and troponin. There are three troponin subunits in vertebrate striated muscle, named differently by each set of workers involved, having molecular weights of about 17,000, 22,000, and 38,000 (Ebashi *et al.*, 1973; Hartshorne, 1973; Greaser and Gergely, 1973; Perry *et al.*, 1973). The two molecules (tropomyosin troponin) interact to effect the calcium-ion control of the interaction of actin and myosin. We will call this two-molecule system the "relaxing system" since in its absence actin and myosin can interact fully, and changes in calcium-ion concentration up to levels of, say, 10 μM, produce no change in the action–myosin interaction.

If the relaxing system is present, then the calcium-ion concentration is critical. At low calcium-ion concentrations (below 10^{-8} M) the association of actin with myosin in the presence of ATP is extremely low. Actomyosin gels prepared with the relaxing system present will not superprecipitate following the addition of ATP for tens of minutes. Instead the gel solubilizes, resulting in a "clear phase" in which the actin and myosin are dissociated. Addition of calcium (about 1 μM) then results in rapid superprecipitation. As discussed in the previous section, it is thought that the effect is brought about by the tropomyosin covering myosin binding sites on the actin in the absence of calcium, but being forced to move, thereby uncovering them, in the presence of calcium.

Bremel and Weber (1972) concluded from work at very low ATP concentrations that the relaxing system works only against myosin–nucleotide complexes. In the presence or absence of Ca^{++} a myosin molecule containing no bound nucleotide can bind to actin, thereby displacing the tropomyosin from its position. Since one tropomyosin molecule spans several actin monomers, this displacement by one nucleotide-free myosin causes myosin binding sites on adjacent actins to be made available to myosin–nucleotide complexes. Thus at ATP concentrations sufficiently low that the myosin is present in both nucleotide-free and nucleotide-bound form, the ATPase activity is high, even in the absence of Ca^{++}.

The role of the other proteins of the sarcomere is not clear: there is no evidence that they play a role in the mechanics or biochemistry of the contraction and it is possible that they are purely structural.

3. *Steady-state ATPase*

It appears therefore that an understanding of contraction requires an understanding of the interaction of actin and myosin. Myosin, in the absence of actin, is an ATPase. Considerable work has been done on the ionic requirements of both myosin and actomyosin ATPase. A slightly naïve, but helpful, way of thinking about the interaction between myosin and ATP is to bear in mind that myosin is negatively charged at normal pH, and that ATP is a tetravalent negative ion. This disfavours the interaction. If, however, ATP chelates a divalent ion, normally Mg^{++}, then the result ($Mg\text{–}ATP^{=}$) is only doubly charged, and the interaction between this substrate and the protein is less unlikely. In any event, the normal substrate for ATP hydrolysis in solution is Mg–ATP (Lymn and Taylor, 1970).

The ATPase activity of pure myosin is very low, with a catalytic site activity of about 0.04 sec^{-1} (Eisenberg and Moos, 1968). (Catalytic site activity is the maximal rate of activity that can be obtained at a single site.) The ATPase site is on the subfragment 1. The ATPase activity of myosin is dependent upon the ATP concentration in the manner expected from Michaelis–Menten kinetics.

Each of the two S_1 heads per myosin molecule has one ATP binding site and one actin binding site (Young, 1967). These sites can act independently of one another in the sense that myosin will bind to actin in the absence of ATP, and ATP can bind to and be hydrolysed by myosin in the absence of actin.

Addition of actin to myosin causes a large increase in ATPase activity, resulting under optimal conditions in a catalytic site activity of actomyosin of about 10 sec^{-1} (Eisenberg and Moos, 1970). Eisenberg and Moos (1968, 1970) have shown that the actin–HMM is considerably more dissociated in the presence of ATP than would be expected from the association constant of the actin–HMM complex in the absence of ATP (Young, 1967). *Thus, not only does the actin have an activating effect on the myosin ATPase, but the ATP has a dissociating effect on the actin–myosin complex.*

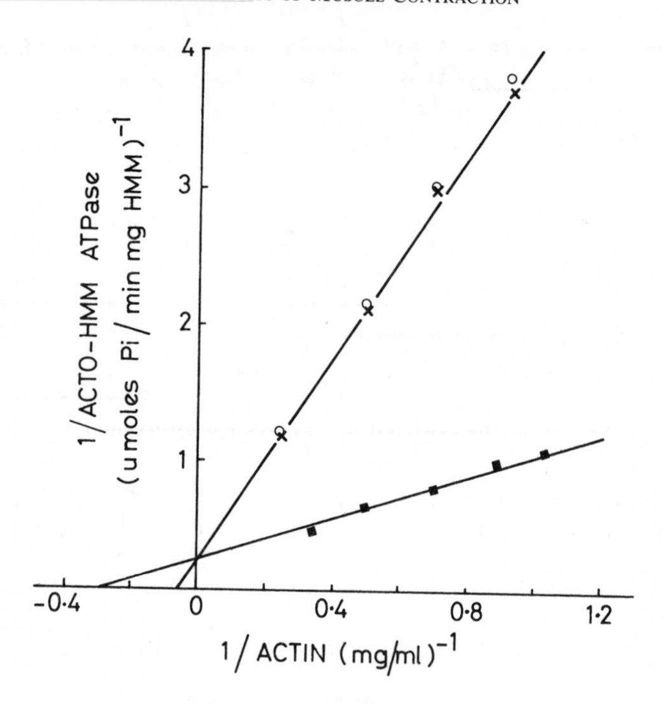

FIG. 15. Reciprocal plot of rabbit actin–heavy meromyosin ATPase as a function of actin concentration. The ionic strengths for curves ○ and × are the same (about 100 mM); ○ denotes an ATP concentration of 2 mM, × an ATP concentration of 0.5 mM. Curve ■: ionic strength about 50 mM, ATP concentration 2 mM. (Redrawn from Eisenberg and Moos, 1968.)

Figure 15, from Eisenberg and Moos (1968), shows, for HMM, the relationship between ATPase activity and actin concentration, plotted on a double reciprocal plot. Three experiments are shown, illustrating the effect of varying the ionic strength and the ATP concentration. Notice that if the ATP concentration is varied, keeping the ionic strength constant, that no change in the ATPase activity results. This is an important result because it shows that the dissociating effect of ATP on the actin–myosin probably occurs by the action of ATP at the hydrolytic site. If there were a separate "dissociating site" and it were not saturated, then the addition of ATP would promote dissociation, upsetting the hydrolysis reaction and resulting in a different relationship between ATPase and actin. This, for example, is what is happening as the ionic strength varies.

Notice the marked effect of ionic strength at these low ionic strengths. The same values of V_{max}, but different slopes, suggests that ionic strength is altering the affinity of the HMM–ATP complex for actin, and therefore that electrostatic interactions are mainly responsible for actin–myosin binding (Rizzino *et al.*, 1970).

The presence of two separate sites on the myosin molecule, one for actin and one for ATP, suggests the following scheme for the various interactions (Eisenberg and Moos, 1968):

However, Moos (1973) has shown that the above scheme cannot account for all the steady-state data. What is needed is some method of investigating individual reactions rather than the integrated effect of a whole network. Techniques for such measurements have been available for a number of years and have recently been applied to muscle proteins.

4. *Rapid-reaction Kinetics*

a. *Myosin ATPase*

In one method to be discussed below, the reactants to be studied are held in separate syringes (Fig. 16 A). The reaction is started by forcing the reactants very rapidly into a reaction chamber. The reaction then starts, and its progress is monitored by some method, usually optically. This is known as a "stopped-flow" experiment, since the flow from the the syringes is stopped before the reaction has proceeded appreciably.

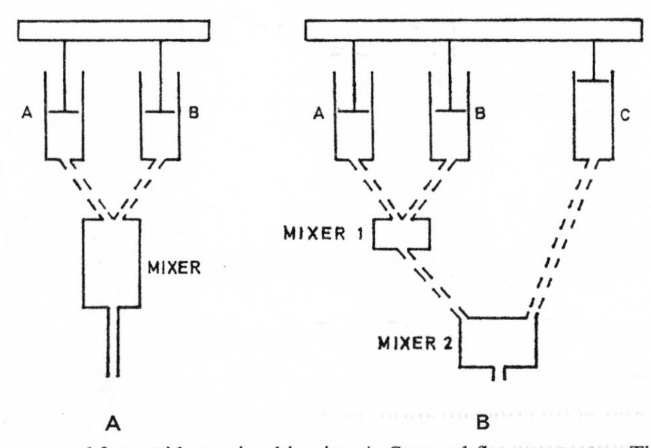

FIG. 16. Apparatus used for rapid-reaction kinetics. A, Stopped-flow apparatus. The reactants are held in syringes *A* and *B*, and are forced into the mixing chamber very rapidly (a few msec). The mixing chamber can be monitored optically in order to measure the progress of the reaction. B, Quenched-flow apparatus. The reactants are held in syringes *A* and *B* as before, and are combined in the small mixing chamber 1. The flow is continuous, and the reaction proceeds in the tube between mixing chambers 1 and 2. The reaction is quenched in mixing chamber 2 by the addition of acid from syringe *C*.

In another type of experiment, known as "quenched-flow", the two syringes holding the reactants feed a small "mixer" chamber (Fig. 16 B). This is connected by a length of tubing to a second mixer chamber. Also with access to this second mixer is a source of strong acid. The reaction between the two reactants takes place in the tubing, and the reaction is stopped by the action of acid. The products can then be analysed. By adjusting the flow rate through the reaction tube or by varying the length of tube, the reaction can be made to proceed for a predetermined time. This "single-point" experiment is laborious because each experiment gives the concentration of products after one preset time interval, the time-course of the reaction requiring a number of such determinations.

Two good introductory accounts of the principles of such measurements are given by Bernhard (1968) and Gutfreund (1965). The time taken to mix the reactants must obviously be much less than the half-time of the reaction being studied. Using motors of a few horsepower to drive syringes of a few millilitres, one can complete the mixing within 1–3 msec.

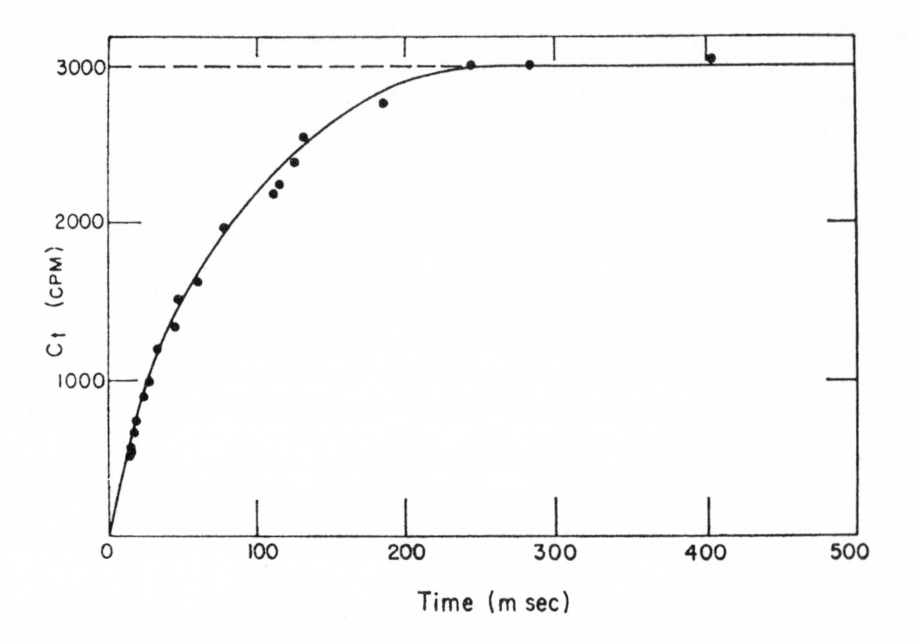

FIG. 17. Time course of phosphate accumulation from quenched-flow experiments in which rabbit myosin was mixed with Mg–ATP (Lymn and Taylor, 1970). The initial ATP concentration was 32 μM. The magnitude of the burst corresponds to 1.2 moles P_i/mole myosin.

The majority of rapid-reaction experiments have been on the hydrolysis of ATP by myosin, HMM, or S_1. First, consider the reaction

$$M + ATP \rightleftharpoons M \cdot ATP \rightleftharpoons M + ADP + P_i + H^+ \tag{2}$$

The steady-state ATPase activity of myosin per catalytic site is known. At an ionic strength of 0.5 it is about 0.02 sec^{-1} (Lymn and Taylor, 1970). The reciprocal of the activity, which we will call the average cycle time, is therefore about 50 sec. For HMM at a low ionic strength the average cycle time is about 20–25 sec (Trentham *et al.*, 1972).

That formula (2) is incomplete was shown conclusively by Lymn and Taylor (1970), who measured the appearance of inorganic phosphate P_i in the acid-quench chamber in a quenched-flow experiment. One syringe contained myosin or HMM and the other Mg-ATP. A series of these experiments, with different flow rates and tube lengths, produces a curve of accumulated P_i "released" vs. time in the course of the reconstructed reaction. The P_i measured of course includes P_i freed in solution *plus* any P_i bound to the myosin at the instant of quenching, but not P_i still incorporated in nucleotide. The result, in Fig. 17, shows that the rate of release of P_i settles down to the value expected from the above steady-state activity of about 0.02 sec^{-1} (the nearly flat slope of P_i vs. time). However, note that there is a rapid early rise of the P_i measured, which accumulates far too rapidly to be accounted for by the steady-state ATPase activity and eqn. (2). Lymn and Taylor have pointed out that one can resolve this anomaly if formula (2) is rewritten

$$M + ATP \rightleftharpoons M \cdot ATP \rightleftharpoons M \cdot ADP \cdot P_i \rightleftharpoons M + ADP + P_i \tag{3}$$

and it is assumed that the rate-determining (slowest) step *follows* the formation of $M \cdot ADP \cdot P_i$; here the $M \cdot ADP \cdot P_i$ can build up rapidly—and the acid quench would detect the P_i in this form.† Once the myosin is mostly in this state, the additional P_i detected represents free P_i at the steady-state rate as expected.

A consequence of this explanation is that the early rise of P_i ought to match quantitatively the amount of myosin injected; it does. However, Trentham *et al.* (1972) were able to demonstrate that formula (3), too, is incomplete. They used a stopped-flow technique and followed the release of either inorganic phosphate or ADP spectroscopically. In this case, without the acid quench, "release" means actual dissociation from the enzyme. In summary (which does not do justice to their extensive and elegant controls), the results are (1) that when HMM and ATP are mixed, a P_i assay shows no initial transient, but only the slow release expected of the steady-state ATPase activity of HMM, (2) that in the same experiment but with an ADP assay, ADP is released at a similar slow rate, and (3) that the HMM–ADP complex itself dissociates at a much more rapid rate (2.3 sec^{-1}) than that corresponding to the ATPase activity. The P_i and ADP releases were measured by linked assay systems. The dissociation of the HMM–ADP complex in (3) was measured via the altered 330 nm absorption of the chromophoric nucleotide thioATP as it occupied myosin sites freed by the dissociation of ADP. The controls on the method will not find space here, but their discussion by Trentham *et al.* (1972), like Lymn and Taylor's (1970) paper, rewards careful reading.

Trentham *et al.* infer from results (1) and (2) (along with several other considerations including the lack of P_i inhibition of ATPase) that the rate-limiting step in formula (3) must precede even the steps constituting the release of both ADP and P_i. We find their inference easier to see if all three results are taken together; for example, since ADP release is clearly rapid from its complex with HMM [result (3)] and yet slow when the initial reactants are HMM and ATP [result (2)], there must be a slow step preceding the occurrence of the HMM–ADP complex. Of course, Lymn and Taylor (above) had already shown that some kind of HMM–ADP complex must form rapidly, *prior* to the rate-limiting step. Thus, provided the releases of P_i and ADP are independent, *two* such states are indicated, connected by the very rate-limiting transition. Trentham *et al.*, therefore, extend formula (3) to read

$$\mathrm{M} + \mathrm{ATP} \underset{k_{-1}}{\overset{k_1}{\rightleftharpoons}} \mathrm{M^*{\cdot}ATP} \underset{k_{-2}}{\overset{k_2}{\rightleftharpoons}} \mathrm{M^*{\cdot}ADP{\cdot}P} \underset{k_{-3}}{\overset{k_3}{\rightleftharpoons}} \mathrm{M{\cdot}ADP{\cdot}P} \underset{k_{-4}}{\overset{k_4}{\rightleftharpoons}} \mathrm{M{\cdot}ADP} + \mathrm{P} \underset{k_{-5}}{\overset{k_5}{\rightleftharpoons}} \mathrm{M} + \mathrm{ADP} + \mathrm{P}, \tag{4a}$$

where $k_3 \simeq 0.04\ sec^{-1}$, k_4 is rapid, and $k_5 \simeq 2.3\ sec^{-1}$ at room temperature and pH 8.

The $\mathrm{M^*{\cdot}ADP{\cdot}P}$ and $\mathrm{M{\cdot}ADP{\cdot}P}$ presumably would differ in conformation or in the manner of the binding of the ligands. Note that if the equilibrium association constant (k_{-5}/k_5) for ADP and myosin is $1.4 \times 10^5\ \mathrm{M}^{-1}$ (Lowey and Luck, 1969), then the above value $k_5 \simeq 2.3\ sec^{-1}$ results in the estimate $k_{-5} = 3.2 \times 10^5\ (\mathrm{M\text{-}sec})^{-1}$.

Note also that the above equation shows the release of phosphate preceding the release of ADP. With our present understanding we should strictly have written

$$\begin{array}{ccccc} & \overset{k_4'}{\nearrow} & \mathrm{M{\cdot}ADP{+}P} & \overset{k_5''}{\searrow} & \\ & \swarrow_{k_{-4}'} & & \nwarrow_{k_{-5}''} & \\ \mathrm{M{\cdot}ADP{\cdot}P} & & & & \mathrm{M{+}ADP{+}P} \\ & \searrow^{k_5'} & & \nearrow^{k_4''} & \\ & \underset{k_{-5}'}{\nwarrow} & \mathrm{M{\cdot}P{+}ADP} & \underset{k_{-4}''}{\swarrow} & \end{array} \tag{4b}$$

† Similar experimental methods have been used for a number of years by Tonomura and his colleagues. Their explanations are much more complex than those of Taylor. The latest paper on the work from Tonomura's laboratory is Inoue *et al.* (1972).

On the other hand, Taylor has pointed out to us (personal communication) that it is possible to explain Trentham's results invoking an alternative slow step provided that the release of P_i is slow and necessarily precedes that of ADP. In the discussions below of the reactions which follow the first two in Formula 4a we have elected to follow Trentham's interpretation. However, discussion of reactions 1 and 2 is independent of these alternatives.

b. *Reactions* 1 *and* 2

As shown above and in Fig. 17, Lymn and Taylor (1970) measured the P_i (associated and dissociated) accumulation as a function of time after addition of ATP to myosin. The rate constant for the exponential saturation of the early accumulation of phosphate in Fig. 17 is 13.8 sec^{-1}. The experiment can now be repeated for different values of initial ATP concentration, and a curve of this rate constant plotted vs. starting ATP concentration. Their result is shown in Fig. 18 A. For low ATP concentrations this rate constant is approximately proportional to the ATP concentration, but saturates at about 125 mM ATP at a value of 50–60 sec^{-1}.

Consider why this is so. We know from the above reasoning that the P_i "released" in this experiment will be bound to the myosin prior to the rate-determining step. Only the forward reaction in step 1 in this scheme is ATP-dependent. When the ATP concentration is high enough, the M·ATP complex will be formed sufficiently rapidly that the limiting rate constant will be $(k_2 + k_{-2})$ (see Appendix I, with the condition that k_3 is very small with respect to $(k_2 + k_{-2})$). We see from Fig. 18 A that for myosin at high ionic strength $(k_2 + k_{-2})$ is about 55 sec^{-1}. Figure 18 B and C, are taken from Lymn and Taylor (1970, 1971) and show that for HMM at low ionic strength $(k_2 + k_{-2})$ is about 160 sec^{-1}.

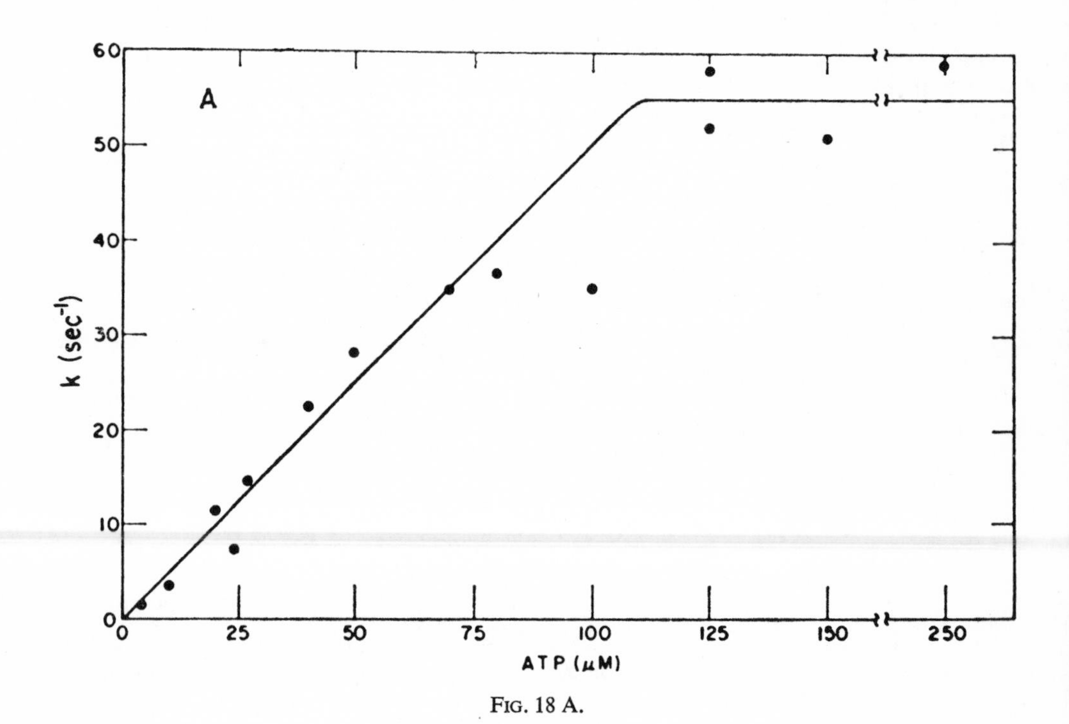

FIG. 18 A.

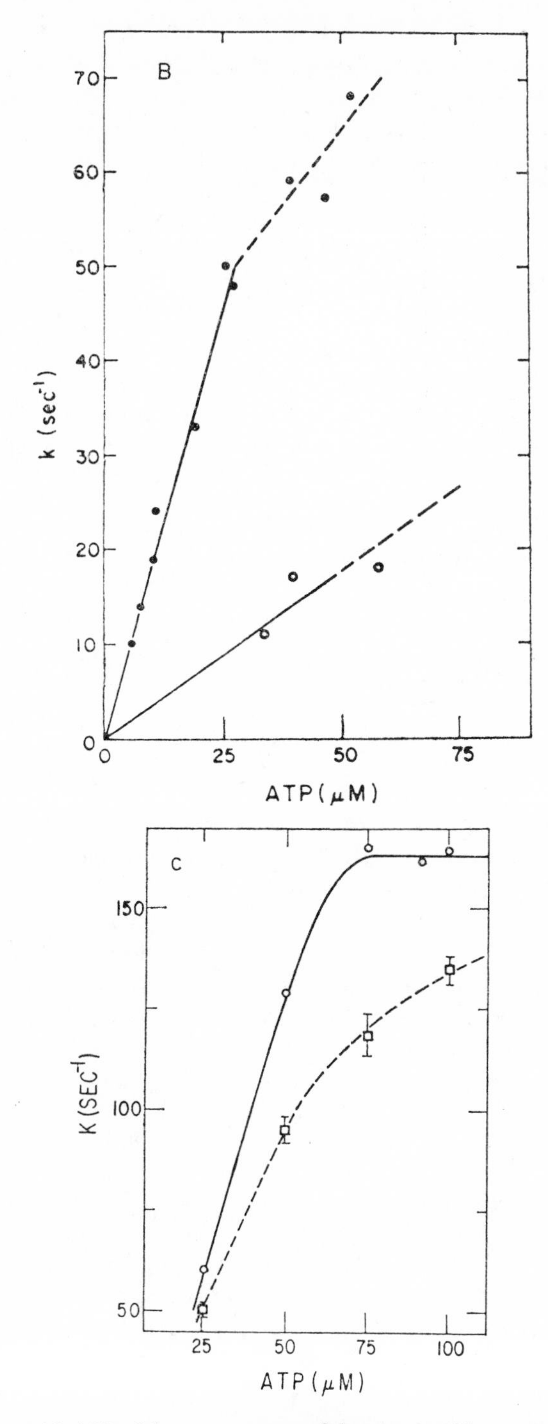

FIG. 18. Variation with ATP of the rate constant of the exponential curve measured in a series of experiments (on rabbit muscle proteins) such as that of Fig. 17. A, With myosin at high ionic strength. B, With HMM at high ○ and low ● ionic strength. C, With HMM ○ and actin–HMM □ at low ionic strength. (A and B from Lymn and Taylor, 1970; C from Lymn and Taylor, 1971.)

Lymn and Taylor (1970, 1971) make the assumption here, and below in formulae (5a) and (5b), that the backward steps are very slow compared with the forward steps, and thus ignore k_{-2} in the above expression. The reversibility of each step must be tested experimentally before this simplification can be justified.

Bagshaw *et al.* (1973) separated k_2 and k_{-2} as follows. They quenched the reaction after 1 sec, i.e., before the reaction $M^* \cdot ADP \cdot P \rightleftharpoons M \cdot ADP \cdot P$ has taken place to any appreciable extent. The ATP ought to be equilibrated between $M^* \cdot ATP$ and $M^* \cdot ADP \cdot P$ (step 2). By this means they obtained an estimate of the equilibrium constant for this reaction. They found, in preliminary experiments, that k_{-2}/k_2 is approximately 0.14; since $k_2 + k_{-2} = 160\ \text{sec}^{-1}$, $k_2 = 140\ \text{sec}^{-1}$ and $k_{-2} = 20\ \text{sec}^{-1}$.

As Lymn and Taylor (1970) point out, there are two interpretations of Fig. 18 at low ATP concentrations that give rise to the observed exponential accumulation of phosphate—either with k_{-1} much greater than or much less than $k_2 + k_{-2}$. These two situations are analysed in Appendix I. For $k_{-1} \ll k_2 + k_{-2}$ the rate constant is

$$k_1 \cdot \text{ATP} + \frac{k_{-1} \cdot k_{-2}}{k_2 + k_{-2}} \tag{5a}$$

and for $k_{-1} \gg (k_2 + k_{-2})$ the rate constant is

$$\frac{k_2}{k_{-1}} \cdot k_1 \cdot \text{ATP} + k_{-2}. \tag{5b}$$

In both cases the rate constant is linearly related to the ATP concentration. Lymn and Taylor's evidence does not distinguish between the two possibilities.

However, Bagshaw *et al.* (1973) have suggested that the first of these is correct.

(1) They measured the change in protein fluorescence when ATP is added to S_1 in a stopped-flow experiment. One can interpret this increase in fluorescence as the sum of $M^* \cdot ATP$ and $M^* \cdot ADP \cdot P$ since they showed that these complexes exhibit identical absolute fluorescence. If k_{-1} is much less than $(k_2 + k_{-2})$, then at high ATP concentrations the fluorescence change will occur as a single exponential (with rate constant $k_1 \cdot \text{ATP} + (k_{-1} \cdot k_{-2})/(k_2 + k_{-2})$; see Appendix I), but if much greater then there will be two exponential phases, the first due to the rapid equilibrium between $M^* \cdot ADP$ and M, and the second as $M^* \cdot ATP$ cleaves to $M^* \cdot ADP \cdot P$, allowing more M to bind ATP in order to maintain the equilibrium of the first reaction. In fact, their results showed only one exponential phase over a wide range of initial ATP concentrations, suggesting that k_{-1} is less than one-tenth $(k_2 + k_{-2})$, i.e. $k_{-1} \ll 160\ \text{sec}^{-1}$.

(2) They also repeated Lymn and Taylor's (1970) quenched-flow experiment, using a concentration of ATP such that $k_1 \cdot \text{ATP} \simeq (k_2 + k_{-2})$. As indicated in Appendix I, the resultant time-course should be sigmoidal for $k_{-1} < (k_2 + k_{-2})$ and a simple exponential for $k_{-1} \gg (k_2 + k_{-2})$. Their result is shown in Fig. 19. Bagshaw *et al.* (1973) point out that the experimental points have large errors, but that their arrangement seems more sigmoidal than exponential, thus tending to support the above conclusion.

To the extent that this argument is correct one can use the first of the above alternatives to interpret the slope and intercept in Lymn and Taylor's exponential phosphate accumulation (Fig. 18). This procedure yields $k_1 = 2.4 \times 10^6\ (\text{M-sec})^{-1}$.

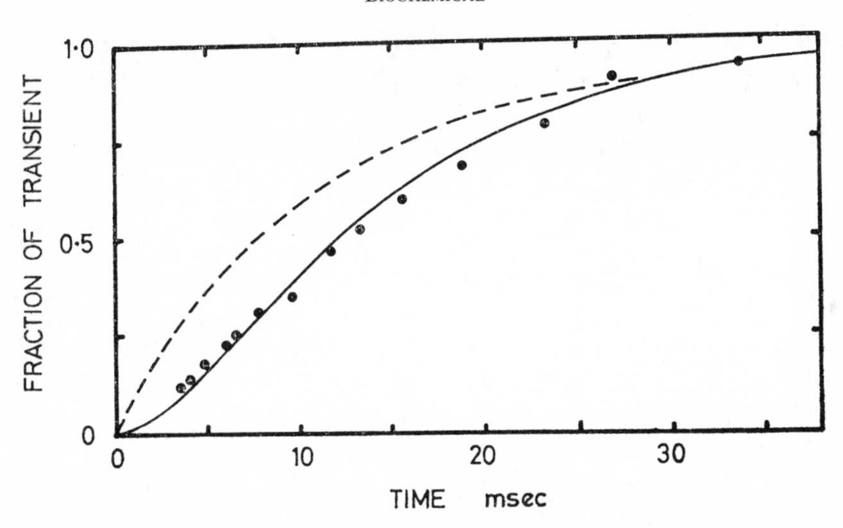

FIG. 19. Time-course of phosphate accumulation in a quenched-flow experiment similar to that of Figs. 17 and 18, but using a concentration of ATP such that $k_1 \cdot \text{ATP} \simeq (k_2 + k_{-2})$ as discussed in the text. The dashed and full lines are the predicted time-courses for the two alternative explanations, as also discussed. 67 μM ATP, 30 μM rabbit S_1 heads. (Redrawn from Bagshaw *et al.*, 1973.)

c. *Actin–Myosin ATPase*

(i) *Kinetics of actin–myosin interaction.* Knowledge of the sequence of steps involved in actin–myosin ATPase is incomplete. We remarked in the section on steady-state biochemistry that there are two competing reactions in the actin–myosin interaction with ATP: (1) actin activates the myosin ATPase, (2) ATP enhances the actin–myosin dissociation. A final kinetic scheme must account for these.

Each of the steps in the scheme for the hydrolysis of ATP by myosin may have a counterpart in which actin is bound to the myosin. Furthermore, there may be exchange between the members of each such pair of states via binding of actin. By these criteria alone, a kinetic scheme is as follows (formula 6):

$$
\begin{array}{ccccccccccc}
\text{A} & & \text{A} & & \text{A} & & \text{A} & & \text{A} & & \text{A} \\
+ & & + & & + & & + & & + & & + \\
\text{M} + \text{ATP} & \underset{k_{-1}}{\overset{k_1}{\rightleftharpoons}} & \text{M*ATP} & \underset{k_{2-}}{\overset{k_2}{\rightleftharpoons}} & \text{M*ADP}\cdot\text{P} & \underset{k_{-3}}{\overset{k_3}{\rightleftharpoons}} & \text{M}\cdot\text{ADP}\cdot\text{P} & \underset{k_{-4}}{\overset{k_4}{\rightleftharpoons}} & \text{M}\cdot\text{ADP} + \text{P} & \underset{k_{-5}}{\overset{k_5}{\rightleftharpoons}} & \text{M} + \text{ADP} + \text{P} \\
kb_{-1} \upharpoonleft\downharpoonright kb_1 & & kb_{-2} \upharpoonleft\downharpoonright kb_2 & & kb_{-3} \upharpoonleft\downharpoonright kb_3 & & kb_{-4} \upharpoonleft\downharpoonright kb_4 & & kb_{-5} \upharpoonleft\downharpoonright kb_5 & & kb_{-6} \upharpoonleft\downharpoonright kb_6 \\
\text{AM} + \text{ATP} & \underset{ka_{-1}}{\overset{ka_1}{\rightleftharpoons}} & \text{AM*ATP} & \underset{ka_{-2}}{\overset{ka_2}{\rightleftharpoons}} & \text{AM*ADP}\cdot\text{P} & \underset{ka_{-3}}{\overset{ka_3}{\rightleftharpoons}} & \text{AM}\cdot\text{ADP}\cdot\text{P} & \underset{ka_{-4}}{\overset{ka_4}{\rightleftharpoons}} & \text{AM}\cdot\text{ADP} + \text{P} & \underset{ka_{-5}}{\overset{ka_5}{\rightleftharpoons}} & \text{AM} + \text{ADP} + \text{P}
\end{array} \quad (6)
$$

Finlayson *et al.* (1969) measured the rate of association (kb_1) of actin and myosin via the turbidity change in a stopped-flow experiment. They report a figure of 1.4×10^5 (M-sec)$^{-1}$.

They also measured the time-course of dissociation of actin–myosin in the presence of Mg–ATP with the same technique, using different concentrations of Mg–ATP. The turbidity change vs. time was well fitted by a single exponential curve, and the rate constant of this curve was proportional to the ATP concentration, with a constant of proportionality in the range 1.4–1.8 × 10^5 (M-sec)$^{-1}$. Finlayson *et al.* point out that there are two plausible schemes resulting in dissociation of actomyosin by ATP:

$$AM + ATP \underset{ka_{-1}}{\overset{ka_1}{\rightleftharpoons}} AM^*ATP \underset{kb_2}{\overset{kb_{-2}}{\rightleftharpoons}} A + M^*ATP \tag{7a}$$

$$AM \underset{kb_1}{\overset{kb_{-1}}{\rightleftharpoons}} A + M \quad M + ATP \underset{k_{-1}}{\overset{k_1}{\rightleftharpoons}} M^*ATP \tag{7b}$$

Scheme (7a) yields a simple exponential turbidity reduction with time constant proportional to ATP concentration only if the rate ($kb_{-2} + kb_2$) of the second reaction is fast compared with that of the first. This scheme corresponds to the conditions of case (2a) in Appendix I, in which case the slope of the relationship between rate constant and ATP concentration gives the value of ka_1 in the above scheme.

Scheme (7b) would yield an exponential decay with rate constant proportional to ATP concentration if the rate-limiting step is the second one, the binding of ATP to myosin. The rate constant of actin–myosin association (above) is 1.4 × 10^5 (M-sec)$^{-1}$. With the actin concentration of approximately 10 μM used in that experiment this amounts to an association rate of 1.4 sec^{-1}. The equilibrium constant of this reaction favours the associated actin–myosin, and thus the dissociation rate (kb_{-1}) will be much slower than 1.4 sec^{-1}. Since the above observed rate constant for the dissociation was 1.6 × 10^5 (M-sec)$^{-1}$, which with 1 mM ATP (within the range used) gives a rate constant of 160 sec^{-1}, it follows that scheme (7b) cannot work.

The experiments have thus given the value of ka_1 in scheme (7a) together with a lower limit on kb_{-2} which must be greater than about 160 sec^{-1}. Lymn and Taylor (1971) repeated the above actin–myosin dissociation experiment with HMM at low ionic strength, and found a value of ka_1 of 1.2 × 10^6 (M-sec)$^{-1}$. They attempted to find a saturation of the measured rate constant by increasing the ATP concentration. They were unable to find a saturating level, and placed a lower limit on kb_{-2} of 1000 sec^{-1}.

Lymn and Taylor (1971) used the quenched-flow technique to measure the rate of phosphate formation for actin–HMM in the manner described earlier for myosin. The results are shown in Fig. 18 C. The same difficulties of interpretation apply, but assuming that the conclusions are also correct for actin–HMM, the value of ka_1 is 1.8 × 10^6 (M-sec)$^{-1}$ measured from the slope of the dashed line in Fig. 18 C. This agrees well with the value deduced above from turbidity measurements. The saturation rate constant was not found because it was not possible to use large enough concentrations of ATP, but the value is probably not much greater than 130–150 sec^{-1}.

Notice that with actin–HMM and ATP there are two pathways by which phosphate can be produced:

$$\begin{array}{ccccc}
AM + ATP & \underset{ka_{-1}}{\overset{ka_1}{\rightleftharpoons}} & AM^*ATP & \underset{ka_{-2}}{\overset{ka_2}{\rightleftharpoons}} & AM^*ADP\cdot P \\
 & & kb_{-2} \downarrow\uparrow kb_2 & & kb_{-3} \downarrow\uparrow kb_3 \\
 & & A + M^*ATP & \underset{k_2}{\overset{k_2}{\rightleftharpoons}} & M^*ADP\cdot P \\
 & & & & + \\
 & & & & A
\end{array} \tag{8}$$

We know that the dissociation rate kb_{-2} is at least 1000 sec^{-1}. kb_2 is much smaller than kb_{-2}. Provided it is sufficiently small, then the kb_{-2} transition will predominate over the

ka_2 transition unless the latter is comparable to 1000 sec^{-1} as well. However, if this were the case the experimental rate constant would not saturate at the relatively low value of 150 sec^{-1}. Therefore the ka_2 transition cannot occur appreciably and it must be the k_2 transition which limits the saturation at 150 sec^{-1}. Moreover, this figure agrees well with that (160 sec^{-1}) found from the experiments with HMM for the k_2 transition.

If the predominant pathway of hydrolysis of ATP by actin–myosin involves the kb_{-2} dissociation of actin–myosin as above, then the question arises as to the mechanism by which actin actually activates the myosin ATPase. We have seen that step 3 of the myosin pathway is the rate-limiting step. One obvious way for actin activation to occur, as pointed out by Lymn and Taylor (1971), is for the rate constant for the binding of actin to myosin in the state before the rate limiting step (i.e. kb_3 in the scheme here) to be greater than that of the rate-limiting step of the myosin ATPase (i.e. k_3). They attempted to measure this value, but their technique (rapid elution through a column) is prone to large errors, and their figure cannot be used with any degree of certainty.

TABLE 2. RATE CONSTANTS OF MYOSIN AND ACTIN–MYOSIN ATPase
The rate constants are identified using the terminology of formula (6).

Rate constant	Value	Units	Myosin species	Temp. (°C)	Ionic strength	pH	Source
k_1	5×10^5	(M-sec)$^{-1}$	M	20	0.5	8.0	LT'70
k_1	7.2×10^4	(M-sec)$^{-1}$	M	0	0.5	8.0	LT'70
k_1	2.4×10^6	(M-sec)$^{-1}$	HMM	20	0.05	8.0	LT'70
k_1	5×10^5	(M-sec)$^{-1}$	HMM	20	0.5	8.0	LT'70
$k_2 + k_{-2}$	55	sec^{-1}	M	20	0.5	8.0	LT'70
$k_2 + k_{-2}$	125	sec^{-1}	HMM	20	0.05	8.0	LT'70
$k_2 + k_{-2}$	160	sec^{-1}	HMM	20	0.05	8.0	LT'71
ka_1	1.8×10^6	(M-sec)$^{-1}$	HMM	20	0.05	8.0	LT'71
$ka_2 + ka_{-2}$	150	sec^{-1}	HMM	20	0.05	8.0	LT'71
kb_1	1.4×10^5	(M-sec)$^{-1}$	M	20	0.5	8.0	FLT'69
kb_{-2}	1000	sec^{-1}	HMM	20	0.05	8.0	LT'71
k_3	0.04	sec^{-1}	HMM	21	0.08	8.0	TBEW'72
k_5	2.3	sec^{-1}	HMM	20	0.2	8.0	TBEW'72
k_{-5}/k_5	1.4×10^5	M^{-1}	M		0.6	7.7	LL'69
k_{-5}/k_5	1×10^5	M^{-1}	HMM		0.15	7.7	LL'69

References: FLT'69: Finlayson *et al.*, 1969. LL'69: Lowey and Luck, 1969. LT'70: Lymn and Taylor 1970. LT'71: Lymn and Taylor, 1971. TBEW'72: Trentham *et al.*, 1972.

Rate constants have not yet been measured for other steps. Figure 20 A and Table 2 summarize the findings, and Fig. 20 B is an adaptation from Lymn and Taylor (1971), taking account of the results from Trentham's laboratory (Trentham *et al.*, 1972; Bagshaw *et al.*, 1973), of the likely steps involved in the hydrolysis of ATP by actin–myosin, and by implication, the steps involved in muscle contraction. Note that the steps of this suggested sequence account for the dual role of ATP mentioned earlier—its function as a substrate and its ability to dissociate actin–myosin—without requiring a second "dissociating" site, which has been suggested many times. It may be precisely because of this ability to dissociate actin–myosin during its hydrolysis that muscle can link its mechanical to its biochemical operation so well. It is difficult to see how there could be effective chemo-mechanical coupling otherwise, with, for example, a second ATP site which caused actin–myosin dissociation.

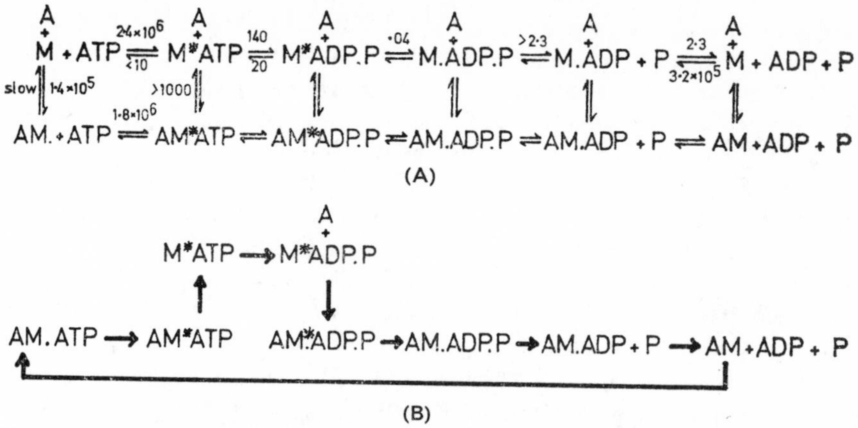

$$\begin{array}{ccccccccccc}
\overset{A}{\overset{+}{M}}+ATP & \underset{<10}{\overset{2.4\times10^{6}}{\rightleftharpoons}} & \overset{A}{\overset{+}{M^{*}ATP}} & \underset{20}{\overset{140}{\rightleftharpoons}} & \overset{A}{\overset{+}{M^{*}ADP.P}} & \overset{.04}{\rightleftharpoons} & \overset{A}{\overset{+}{M.ADP.P}} & \overset{>2.3}{\rightleftharpoons} & \overset{A}{\overset{+}{M.ADP}}+P & \underset{3.2\times10^{5}}{\overset{2.3}{\rightleftharpoons}} & \overset{A}{\overset{+}{M}}+ADP+P \\
\text{slow}\updownarrow 1.4\times10^{5} & & >1000 \updownarrow 1.8\times10^{6} & & \updownarrow & & \updownarrow & & \updownarrow & & \updownarrow \\
AM.+ATP & \rightleftharpoons & AM^{*}ATP & \rightleftharpoons & AM^{*}ADP.P & \rightleftharpoons & AM.ADP.P & \rightleftharpoons & AM.ADP+P & \rightleftharpoons & AM+ADP+P
\end{array}$$

(A)

$$\begin{array}{ccccc}
 & M^{*}ATP & \rightarrow & \overset{A}{\overset{+}{M^{*}ADP.P}} & \\
 & \uparrow & & \downarrow & \\
AM.ATP \rightarrow & AM^{*}ATP & & AM^{*}ADP.P \rightarrow AM.ADP.P \rightarrow AM.ADP+P \rightarrow AM+ADP+P &
\end{array}$$

(with an arrow returning from AM + ADP + P to AM.ATP)

(B)

FIG. 20. A, Summary of the reactions and rate constants discussed in the text. B, Suggested most probable sequence of reactions in actively contracting muscle.

(ii) *Steady-state actin–myosin interaction in the presence of ATP.* Recent work by Eisenberg *et al.* (1972) and Eisenberg and Kielley (1973) has provided further information about the steps involved in the hydrolysis of ATP by actomyosin. They have investigated thoroughly the phenomenon first reported by Leadbetter and Perry (1963) that under normal conditions of high ATPase activity by mixtures of actin and HMM, the viscosity of the solution is much lower than would be the case if the actin and the HMM were fully associated. One implication of Leadbetter and Perry's result is that for a significant portion of the cycle of hydrolysis of ATP the actin and myosin are separate. In order to investigate this idea, Eisenberg *et al.* (1972) and Eisenberg and Kielley (1973) measured the rate of sedimentation of actin and either HMM or S_1 in an analytical ultracentrifuge—in both the presence and absence of ATP. By this means they were able to determine the percentage of the HMM or S_1 which was not bound to the actin, since this myosin sedimented at a slower rate. In the absence of ATP, the HMM sedimented with the actin, suggesting that it was all bound, as expected. In the presence of ATP, under conditions in which the ATPase activity was about 80% of its maximal value (as determined from Lineweaver-Burke plots), the HMM was only just over 40% bound and the S_1 about 20% bound. Eisenberg and Kielley interpret their data to suggest that under conditions of maximal activity only about 25% of the S_1 is bound to the actin. Note that this interpretation depends critically upon the estimate of the maximal activity of the actin–myosin ATPase.

That their results were not due to the effect of centrifugation was confirmed by measuring the ATPase activity of the system first as a function of actin concentration, and then as a function of HMM or S_1 concentration. Figure 21 shows a set of results for S_1 in which the ATPase activity is plotted in units of activity per S_1 monomer (when actin is varied) or activity per actin monomer (when S_1 is varied). The maximum activity obtainable from each S_1 is found to be much less than that for each actin. This suggests that several S_1's can interact with one actin in the average time that one S_1 completes its cycle; one can rule out the possibility that actin is bound to the S_1 throughout its cycle of activity. These results would be much more convincing if they also conveyed some information about the variance in the measurements.

On the basis of these results one can rule out the initially attractive suggestion that the rate of actin–myosin association (which under the conditions of this experiment is kb_3) is rate

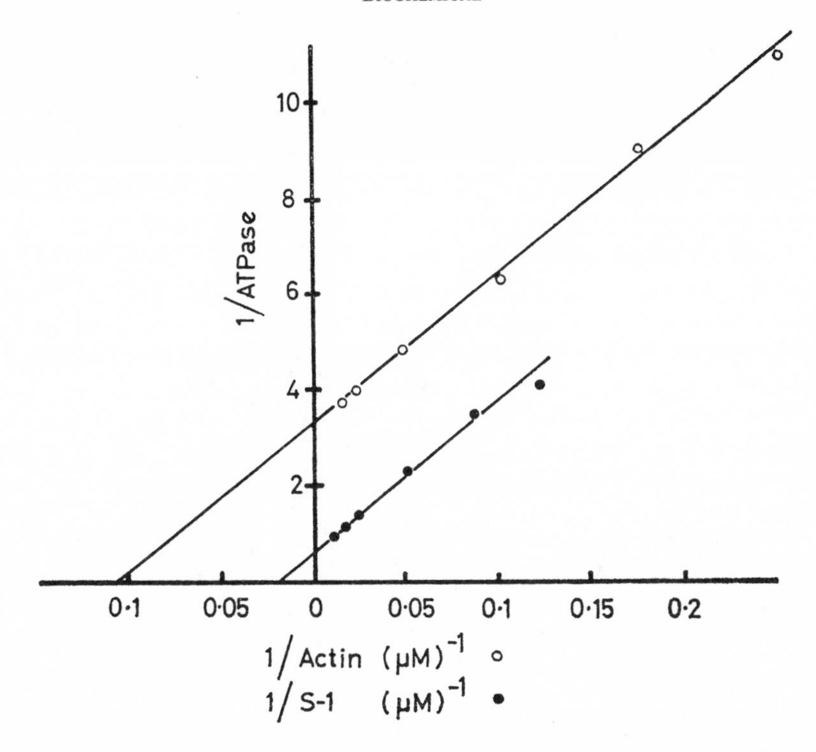

FIG. 21. Reciprocal activity plots of rabbit S_1 ATPase plotted ○ as a function of the actin concentration and ● as a function of the S_1 concentration. The ATPase activity is plotted as μM phosphate released per μM actin per second for the plot vs. $1/S_1$, and as μM phosphate released per μM S_1 per second for the plot vs. 1/actin. The intercepts on the 1/ATPase axis thus compare the maximal activity per actin monomer at infinite S_1 with the maximal activity per S_1 monomer at infinite actin. (Redrawn from Eisenberg and Kielley, 1973.)

limiting; kb_3 is a bimolecular rate constant, and thus, if the actin concentration is made high enough, can be very large. The above interpretation uses values extrapolated to infinite actin concentrations, and thus cannot have kb_3 as the rate limiting step. For this reason Eisenberg and Kielley (1973) suggest an intermediate "refractory" state for the detached M·ADP·P complex. To account for the observation that about 75% of the S_1 is detached from actin at a maximal ATPase activity of about 10 sec^{-1}, the rate constant for the reaction following this refractory state must be of the order of 14 sec^{-1}. Such a refractory state cannot therefore be any of the complexes included in formula (6) (k_2 is much faster, and k_3 much slower, than 14 sec^{-1}). Of course, it remains of considerable interest to know the value of kb_3, which might, for example, be measured in a double stopped-flow experiment. (HMM and ATP are first mixed; a short time later actin is added, and the resulting optical density changes measured.)

The lack of association found by Eisenberg *et al.* (1972) and Eisenberg and Kielley (1973) fits in well, as they point out, with the X-ray diffraction data discussed earlier which suggest that under conditions of high activity in intact muscle only about 20–30% of the cross-bridges are attached.

Finally, one can place limits on the values of the rate constants for the actin–myosin attached states in our working scheme, assuming that the myosin is detached for about

75% of the cycle time. The sum of the reciprocal values of the rate constants ($1/ka_3 + 1/ka_4 + 1/ka_5$) is then at least 20 msec, and thus none of these three rate constants can be less than 50 sec^{-1}. However, steps 3 and 5 in the myosin pathway are 0.04 sec^{-1} and 2.3 sec^{-1}, respectively. It is necessary, therefore, that with actin bound both of these steps are accelerated.

V. HIGH-TIME-RESOLUTION MECHANICAL EXPERIMENTS

Here we describe current methods of measurement and description of mechanical transients, as a prelude to their interpretation in § VI.

Jewell and Wilkie's experiments (1958), discussed in § III, demonstrated that muscle could not be described by a two-element model with non-interacting and time-independent properties. One reason they were able to do this was that they took great trouble to eliminate compliance in their apparatus and attachments to the muscle. The more such compliance, the more closely the entire assembly matches Hill's formulation, and hence the more difficult it is to exclude the model.

Their experiments also suggested that one ought to try, with improved time resolution, to see what is happening during the period immediately following the step change in tension. A necessary conclusion from their results is that the extrapolation they had to make in this period (based upon the validity of Hill's model) is incorrect. First, it is desirable to make the duration of "step" perturbations (actually finite ramps) as short as possible. At the moment, due largely to the work of A. F. Huxley and his collaborators, this time is slightly less than 1 msec for the application of either step changes in tension, or step changes in length. Second, the tension or length transducer must follow the resultant muscle response with a similar resolution. A. F. Huxley's work has set the standard here, too; transducers with resonant frequencies over 4000 Hz are now used.

If there is any series elasticity in muscle which is not part of the contractile mechanism, then step tension changes have one advantage over step length changes: length changes in the series element occur *pari passu* with the applied tension changes, and then remain constant. On the other hand, when tension fluctuates following the application of a step change in length, then the length of any series elastic elements will also change, resulting in length changes of those structures responsible for the contraction. Until recently it has been thought that there were significant series elastic elements in muscle, and therefore step tension changes were favoured. We will describe the results from these first.

Civan and Podolsky (1966) extended the work of Podolsky (1960), who had improved the time-resolution of his apparatus to make it faster than that of Jewell and Wilkie (1958). The experiments were in principle the same as the isotonic-release experiments of Jewell and Wilkie, i.e. after development of an isometric tetanus the muscle was suddenly released and then allowed to shorten against a constant force. Figure 22 shows Civan and Podolsky's (1966) results. The first two records are on a slow time scale, showing the time course of a normal isometric tetanus and the effect of applying a step change in tension. The other records, on a faster time scale, show the response of the muscle to tension changes of different amplitudes. Notice that there is still early oscillation, attributed to the apparatus. But this now disappears within about 2–3 msec, and is of a fairly small amplitude compared with the size of the overall change in tension.

The dashed lines show the extrapolation from the region of constant velocity which is reached, in these records, after 10–50 msec depending upon the size of the step. There are

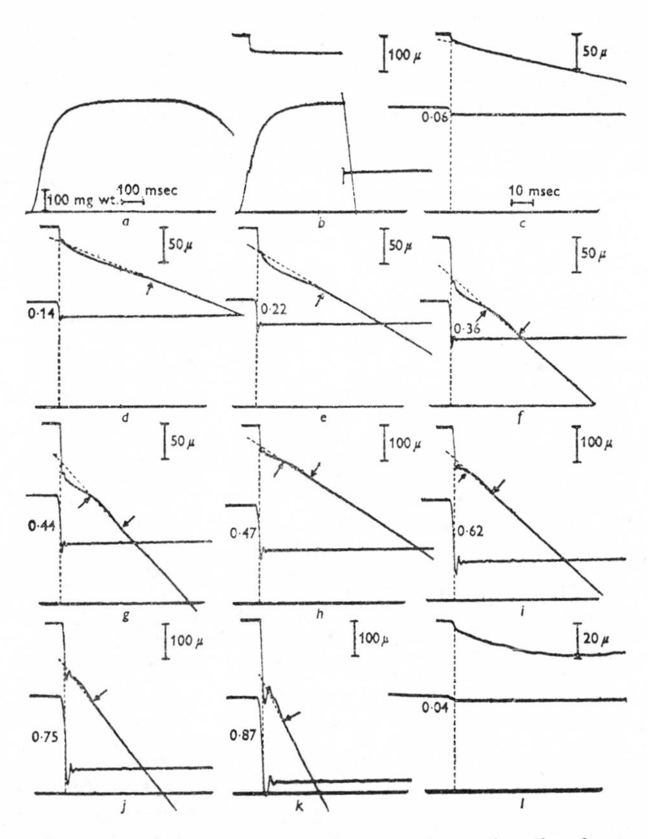

FIG. 22. Tension–step (isotonic-release) experiments using a bundle of seven frog semi-tendinosus fibres at 3°C. *a* and *b* are on a slow time scale to illustrate the general form of the experiments; *c–l* are on a faster time scale; each one has both the length (upper) and the tension (lower) record. The figures alongside each record give the final value of the tension as a fraction of the full isometric tension. The arrows mark the null points of the oscillations. (From Civan and Podolsky, 1966.)

obvious and significant departures from the constant-velocity line. In very rough terms the response approaches constant velocity via damped oscillations (at a frequency much below that of the above spurious oscillations). For small step-tension changes the response might be described as being so heavily damped that no oscillations are seen; the minimum damping seems to occur when the tension is changed by about 40% of its initial value.

Figure 23, parts a and b, shows a similar experiment by Armstrong *et al.* (1966). Here, the response to both a small step increase and step decrease in length is shown. The response to a step decrease resembles that of Civan and Podolsky. When the tension is increased suddenly by about 10%, however, the oscillations in length are much more pronounced. The tension was maintained at the required level by a feedback mechanism. (Tension was measured at one end of the suspended fibre. The tension signal, in the form of a voltage, was compared with a controlled reference voltage. The difference between the two was used to drive the length transducer to apply compensatory length changes at the other end of the fibre.) The length changes in the central region were monitored by spot followers (Gordon *et al.*, 1966). There are two advantages of this apparatus over the balance-arm/

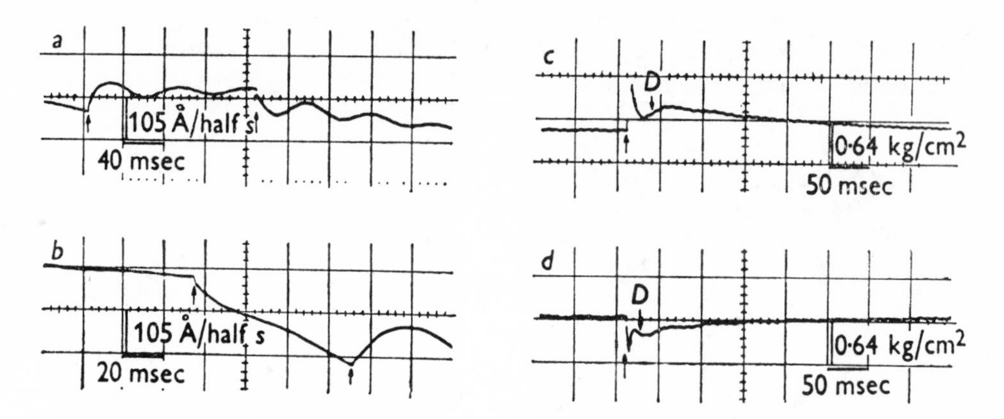

FIG. 23. Mechanical experiments, using the spot-follower, on single frog semitendinosus muscle fibres at 3°C. *a* and *b*, length records from tension–step experiments. *a*, The tension was increased from 2.89 to 3.24 kg/cm² at the first arrow and returned to 2.91 kg/cm² at the second arrow. Isometric tension 3.25 kg/cm². *b*, The tension was decreased from 2.95 to 2.66 kg/cm² at the first arrow and returned to 2.95 kg/cm² at the second arrow. Isometric tension 3.25 kg/cm². Note that in both records the responses are superimposed on a steady shortening of the fibres. *c* and *d*, Tension records from length–step experiments. In *c* the response is shown to a step increase of 4.8 nm per half-sarcomere, and in *d* to a decrease of 4.6 nm per half-sarcomere. *D* indicates the delayed tension change. (From Armstrong *et al.*, 1966.)

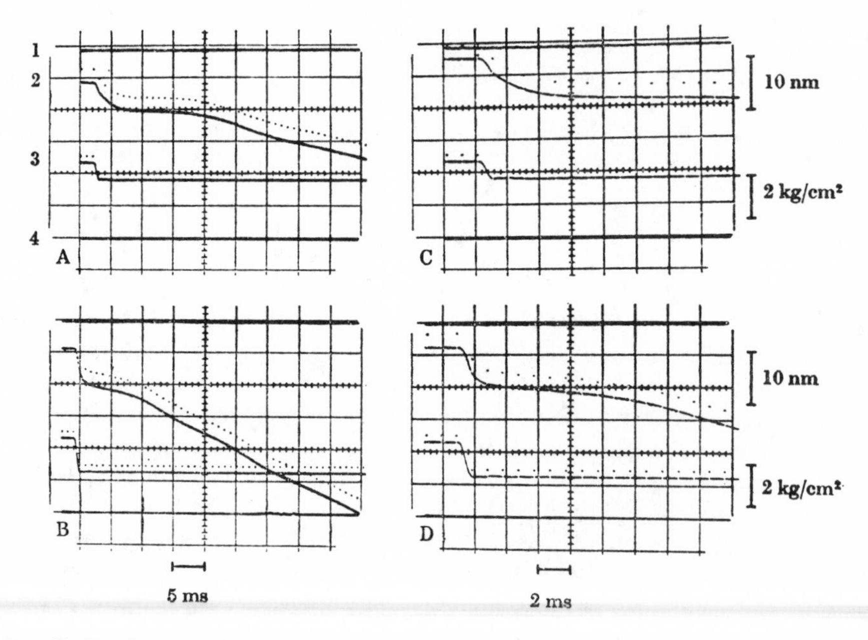

FIG. 24. Tension–step experiments, using the spot-follower, on single frog semitendinosus fibres at 2.5°C. Trace 1, baseline for length (corresponding to sarcomere length of 2.14 μm); trace 2, length record (shortening downwards); trace 3, tension record; trace 4, baseline for tension. The dots above the records are at 1 msec intervals. In A and C the tension is lowered by 25%, in B and D by 44%. (Experiment of A. F. Huxley and Simmons, from A. F. Huxley, 1971.)

spring method of others: (a) all effects of extraneous series elasticity are eliminated, together with those of the known non-uniformity of sarcomeres near the tendon, (b) inertial forces are reduced to a minimum and, since the step change in tension is controlled by the waveform of the reference level, one can compensate to some extent for mechanical imperfections in the apparatus.

More recent experiments by A. F. Huxley and Simmons (from A. F. Huxley, 1971) are shown in Fig. 24. Tension is changed in less than 1 msec, and no spurious oscillations appear. This remarkable improvement required the development of a new tension transducer (A. F. Huxley and Simmons, 1968) as well as the use of the spot follower. A. F. Huxley and Simmons (1973) discuss the difficulties of making such measurements. The main problem

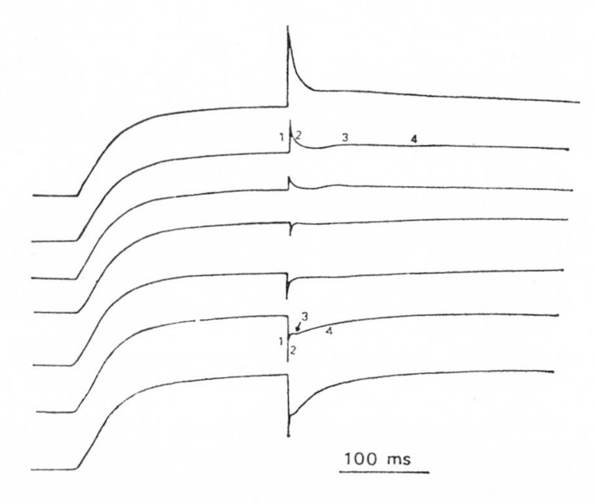

FIG. 25. Length–step experiments on a slow time scale, using single frog semitendinosus muscle fibres. The rise in tension during tetanization is shown, followed by the response to a step length-change. The largest step increase is 5.4 nm per half-sarcomere, and the largest step decrease is 8.8 nm per half-sarcomere. (The records from A. F. Huxley and Simmons, 1973, are retouched tracings.)

is that the nonlinear properties of the muscle fibre influence the length changes necessary to maintain the tension, and these have to be taken into account in the electronics of the feedback mechanism. Since the muscle properties change with step magnitude, greatly complicating the design, they altered their procedure to apply step length changes, for which this difficulty does not arise.

Two preliminary records of responses to small step changes in length were shown in Fig. 23, parts c and d, from Armstrong *et al.* (1966). There is an initial rapid tension change which occurs as the length is applied, followed by an immediate recovery toward the original isometric level. The rate of recovery then slows markedly, and in the traces in Fig. 23 reverses, so that there is a delayed tension rise in response to a step increase of length, and a delayed tension fall in response to a step decrease in length. The tension then returns to the original isometric tension.

Figure 25 shows a more complete set of records from this kind of experiment, on a fairly slow time scale, from A. F. Huxley and Simmons (1973), and Fig. 26 another set, from an experiment by A. F. Huxley and Simmons (A. F. Huxley, 1971), on a faster time scale.

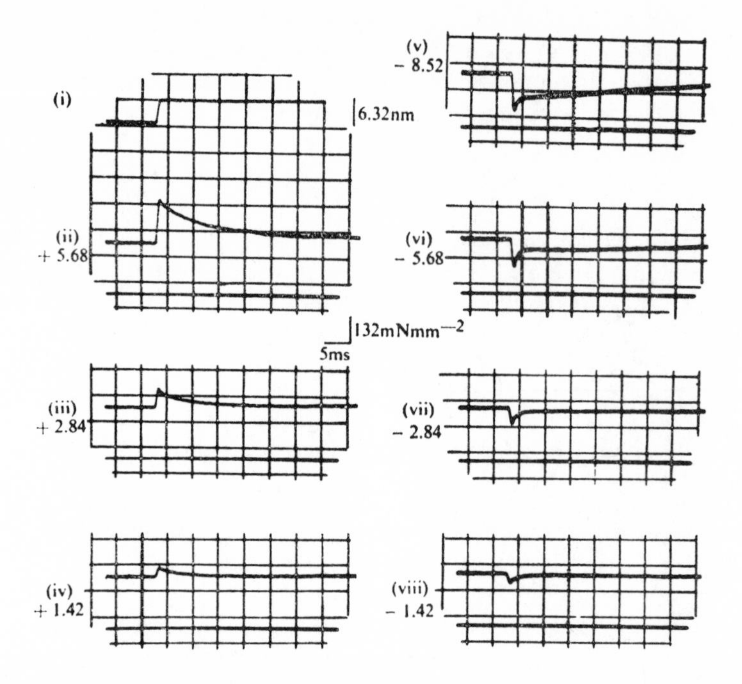

FIG. 26. Length–step experiments on a faster time scale. Upper trace is the length record; middle trace, tension; bottom trace, baseline for the tension record. The figures adjacent to each trace indicate the magnitude of the step length-change in nanometres per half-sarcomere. (From A. F. Huxley and Simmons, 1972.)

The general form of the initial part of any individual response is similar to that expected for a pure elastic element in series with a damped one. Such a structure is shown in Fig. 27, along with the corresponding responses to step changes of both length and tension. To facilitate comparison with A. F. Huxley and Simmons' interpretation of their data (see § VI), the response is shown superimposed upon a starting length L_0, at which length the structure exerts a tension T_0.

With a step decrease in the length of the structure (Fig. 27 B) tension drops instantaneously, to a level T_1. Since a viscous element cannot change length instantaneously the initial length change must have occurred entirely in element k_1. This initial tension change is therefore a measure of the stiffness of this element. Of course, the tension in k_1 must at all times be equal to the sum of the tensions in k_2 and η_2. Since the length of the element k_2 has not yet changed, it must still be exerting its original tension, which is opposed by the viscous force due to η_2. After the initial change there will be a redistribution of length within the structure even though the external length remains constant. k_2 will shorten, extending k_1 until a new equilibrium is reached at which state the new tension T_2 is determined by the combined properties of k_1 and k_2.

The time-course of the tension change from T_1 to T_2 is exponential, with a time constant $\eta_2/(k_1+k_2)$. If steps of different sizes are applied from the initial length, then T_1 and T_2 will vary in a linear fashion if the springs are linear, and the time constant of the exponential change will be constant. T_1 and T_2 are plotted against step height in Fig. 27 D. Note that T_1 gives the stiffness of k_1, but that T_2 gives the combined stiffness of k_1 and k_2 in series.

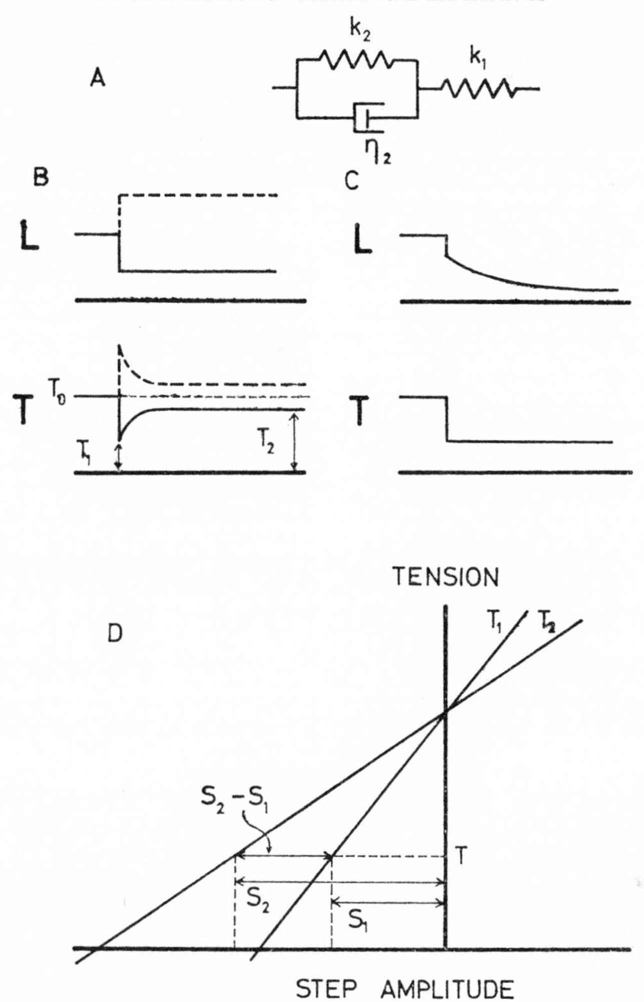

FIG. 27. Linear model used to illustrate Huxley and Simmons' analysis of their length–step experiments. A, The model: a damped spring in series with an undamped spring. B, Response of the model to length–step experiments, starting from an initial extended length. C, Response to a tension–step experiment. D, Analysis of length–step experiments, as discussed in the text.

To determine k_2, one can subtract the step length change (s_1 in Fig. 27 D) which would result in any tension T on the T_1 curve from that (s_2) which would result in the same tension on the T_2 curve. k_2 is $T/(s_2 - s_1)$.

The structure of Fig. 27 produces length changes following step changes in tension like those in Fig. 27 C. In this case the time constant of the exponential portion of the length–time curve is given by η_2/k_2, since the length of the spring k_1 is fixed.

With this general idea in mind, A. F. Huxley and Simmons (1971a) plotted curves of estimated analogous values for T_1 and T_2 vs. length-step height for their data. Naturally, since the time-course in the data is more complex than for the theoretical element considered above, one cannot obtain a complete description in this way. In particular, the "steady level" T_2 is not observed in many cases. Huxley and Simmons assign T_2 at a tension where

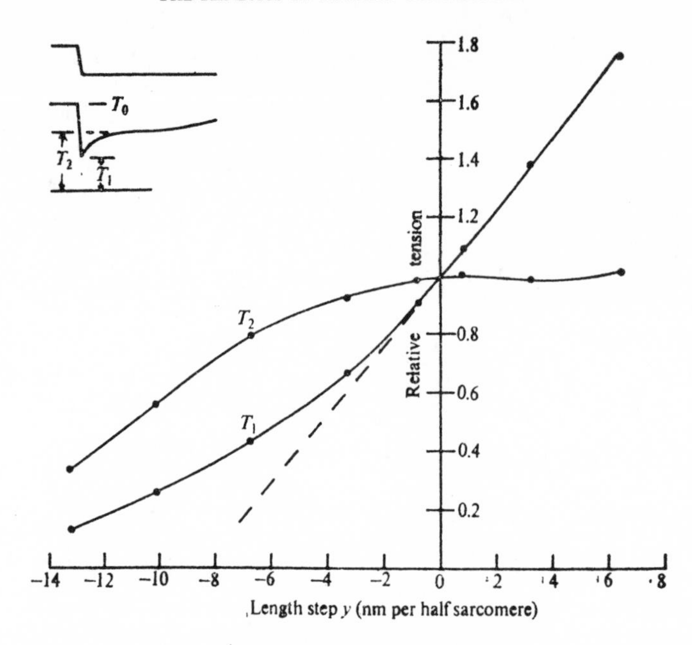

FIG. 28. Analysis of length–step experiments on frog muscle similar to those of Figs. 25 and 26, as discussed in the text and Fig. 27. (From A. F. Huxley and Simmons, 1971b.)

the early relaxation has given way to an inflection or to slower subsequent changes of tension, assume that the latter are due to a separate process to be ignored for the moment and treat only the early events. (We shall see in the next section that Podolsky and Nolan (1972) have devised a model that does not separate the initial from the subsequent events in this manner.) Figure 28 illustrates the form of the T_1 and T_2 curves obtained experimentally.

Whereas the time-course of the tension between T_1 and T_2 in the model of Fig. 27 is exponential, in the case of the muscle this is not usually observed (A. F. Huxley and Simmons, 1971b). However, the time-course of the transition can be expressed as a function of a single "rate constant" of a more complex waveform (A. F. Huxley and Simmons, 1972) and the dependence of this "rate constant" upon step magnitude and direction can be determined experimentally (we have written "rate constant" in quotes to emphasize that this is not the rate constant of a simple exponential process). The "rate constant" is low for step increases of length, but increases rapidly with the magnitude of step decreases of length. We discuss in the next section a stimulating suggestion by Huxley and Simmons for the way this nonlinearity could occur in terms of cross-bridge physics.

Abbott (1972) has suggested that the data of A. F. Huxley and Simmons (1971b) is well described as the sum of two exponential processes of fixed time constant but of varying relative contributions for the different step heights. He goes on to point out that such a description *may* fit in conveniently with the behaviour of the rate constant at very large releases, where the simplest form of the Huxley–Simmons physical hypothesis (next section) calls for large values. Improved time resolution in future measurements will thus be informative. A. F. Huxley and Simmons (1972) suggest that Abbott's description is not satisfactory. They base this view on descriptions which assume that the value of T_2 they measure is, in fact, the final equilibrium value of the transition. However, the value of T_2 measured could be affected by the subsequent processes, ignored in this analysis.

We need not be worried that the above transients are not simple exponentials. For example, any factor which results in a nonuniform conformation of attached cross-bridges (such as filament compliance—see § VI.4—or variation due to non-alignment of A and I filament sites) will produce more complex relaxation. Further, most large-signal stress-relaxations are complicated, if not power-law, decays (Thorson and Biederman-Thorson, 1974).

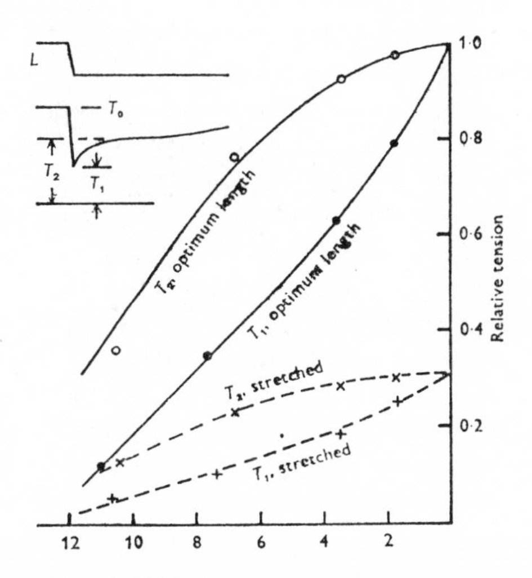

FIG. 29. Analysis of length–step experiments on frog semitendinosus muscle as in Fig. 28 at two different sarcomere lengths: optimum length, 2.2 μm; stretched 3.2 μm. The solid lines are drawn through the circles. The dashed lines are the solid lines scaled down in proportion to the full isometric tensions at the two sarcomere lengths. (From A. F. Huxley and Simmons, 1971a.)

Before proceeding with cross-bridge theory, we shall discuss the current evidence that such experiments in fact measure cross-bridge properties. Figure 28 shows the T_1 and T_2 curves plotted as above from an experiment that included stretches and releases. If the structures responsible for these curves are within the cross-bridges, and if all other structures are effectively rigid, then the magnitude of the response should be proportional to the number of cross-bridges involved, and thus proportional to the amount of overlap. Figure 29 shows T_1 and T_2 curves plotted at two different sarcomere lengths, 2.2 μm and 3.2 μm. The dashed lines are drawn as scaled-down traces of the solid lines in the ratio of the isometric tensions. These are obviously a very good approximation of the experimental points, and show that the hypothesis that the cross-bridges are the structures responsible for the curves is sufficient.

It is difficult to see how any other series elastic element in the sarcomere (such as the Z line or the non-overlapping portions of the filaments) can give rise to these curves since, as A. F. Huxley and Simmons (1973) indicate, the T_1 curve ought then to have a slope independent of tension which, from Fig. 29, is obviously not the case. Moreover, the T_1 and T_2 curves have been shown to scale with tension during the development of tension in an isometric tetanus (A. F. Huxley and Simmons, 1971a, 1973); this, as A. F. Huxley has pointed out, is difficult to explain in terms of a series elastic element.

In summary, no sufficient alternative has yet been offered to the very attractive suggestion that the experiment of Fig. 29 reflects primarily the cross-bridge properties, and we shall follow Huxley's and Simmons's arguments further in the next section.

VI. THEORIES OF CONTRACTION IN MUSCLE

1. *General Comments*

Several analyses have been published which try to relate plausible cross-bridge properties to the behaviour of muscle. The first of these, by A. F. Huxley (1957), was written before the cross-bridges had been seen in electron micrographs (H. E. Huxley, 1957), and has served as the basis for most of the subsequent developments.

The fundamental idea in these formulations is that a cross-bridge has at least two states —(1) attached to the I filament, and (2) detached from the I filament—and that during contraction the cross-bridge goes through these states cyclically. The processes of attachment and detachment are supposed to be biochemical reactions, governed by probabilities per unit time for each transition. Attachment presumably occurs between a site on the cross-bridge and a site on the actin monomer of the I filament.

Consider first the isometric case. Unfortunately the relevant steric constraints are not clear. If no more than one actin monomer per half turn of the actin helix is correctly oriented for binding to a particular cross-bridge (i.e., one actin per 38 nm) then on structural grounds no cross-bridge would be expected to have more than one actin available for binding. If, however, the cross-bridge is sufficiently flexible, then several actins may be available to each cross-bridge. The multiplicity of cross-bridges arising from one site on the backbone of the A filament complicates the issue further, but with over twice as many actin monomers as myosin molecules per sarcomere in vertebrate striated muscle the possibility clearly arises that with sufficient flexibility each cross-bridge has a choice of actins for attachment, even in the isometric case.

Consider for the moment that each bridge has a certain probability per unit time p that it will attach to one of the one or more actin sites in its neighbourhood. Assume that this probability accommodates all factors such as steric constraints and their distribution over the ensemble, thermal motion about these constraints and the actual rate of bond formation given optimum configuration. One of Huxley's plausible propositions was that reactions at these sites are independent; despite the similarity of the equations, one is therefore not dealing with solution-chemistry reactions in which, for example, the rate of binding at a given site can depend upon the free concentration of a reactant. Rather, one computes the probability n that any bridge in the ensemble of independent processes will be attached (the "fraction attached" n) by writing a balance of the attachment and detachment rates for dn/dt. The distinction is clearly seen for the case in which the number of free bridges B equals the number of free actin sites A, for in solution the forward reaction would be bimolecular,

$$\frac{dn}{dt} = pBA = p(1-n)^2, \tag{9a}$$

whereas in the present case it properly retains the simpler monomolecular form

$$\frac{dn}{dt} = pA = pB = p(1-n). \tag{9b}$$

Hence the importance of the relative counts of bridges and sites lies not in dictating the degree of the reaction but rather in determining the physical implication of the number p. Moreover, the choice of normalization (e.g. whether $n =$ fraction of actins or fraction of bridges) must be done with these counts in mind, for if $n = 1$ implies that *all* actins have bridges attached, and there were in fact more such actins than bridges, one could encounter trouble. The case with actins and bridges reversed is similarly a problem. In fact, we (Thorson and White, 1969; White and Thorson, 1972) and Descherevsky (1968) have assigned n to the bridges whereas A. F. Huxley (1957) and those interpreting his formulation (Podolsky and Nolan, 1972, 1973; Julian, 1969) have counted detached and attached actins, defining corresponding rate constants for the attachment of actin to cross-bridges.

Note now what happens with these alternatives when the definition of n is refined to treat not just the total fraction $n(t)$ of attachments at t, but to $n(x,t)$, the fraction attached at time t having distortion x (defined below). Where n is referred to actin sites, the normalization is implicit since the actin sites are *themselves* arrayed uniformly in distortion space, so that $1 - n(x, t)$ is physically the unattached fraction of actin sites at x, where x is, of course, referred to the (common) cross-bridge origin site for the ensemble considered. With n referred to bridges as in our method, however, the conservation of bridges

$$\int_{-\infty}^{\infty} n(x, t)\, dx = 1 - \text{fraction detached}$$

is used explicitly. In either case, the obvious physical implications of allowing n to take on a particular range of values must be considered carefully.

Cross-bridge models have assumed that tension is generated by attached cross-bridges and not by detached cross-bridges. Production of interfilament shear force by a bridge *tends* to cause the I filament to slide past the A filament, so as to produce a shortening of the sarcomere. Whether sliding actually occurs depends upon the total contribution from all other attached cross-bridges and the load on the muscle. The action of sliding will cause the site of attachment of the cross-bridge on the I filament to move past the site of origin of the cross bridge on the A filament. An attached cross-bridge will thus be "distorted" by the action of sliding. The variable "distortion" is important because we expect that it will influence both the force generated by the bridge and the probabilities per unit time of its subsequent changes of state. It is convenient to define distortion in terms of the relative position of the site of origination of the bridge on the A filament and the actin binding site on the head of the cross-bridge, taking the zero of distortion at some convenient relative position. A. F. Huxley (1957) in effect defined the origin of the distortion axis as that point at which the attached cross-bridge generates zero tension.

2. *A. F. Huxley's* 1957 *Formulation and its Exegeses*

A. F. Huxley (1957) proposed that contractile force comprises the interfilament shear forces due to independent sites of interaction between the thick and thin filaments. "Side-pieces" (or "cross-bridges"—H. E. Huxley, 1957), coupled elastically to the thick filaments, were considered to contribute cyclically to this force. In this way, local molecular events could be related to the gross filament sliding required by the sliding-filament idea.

To characterize these cross-bridge cycles, Huxley wrote a differential equation for the balance of the making and breaking of attachments by bridges at thin-filament sites:

$$\frac{\partial n(x, t)}{\partial t} = f(x)[1 - n(x, t)] - g(x)n(x, t). \tag{10}$$

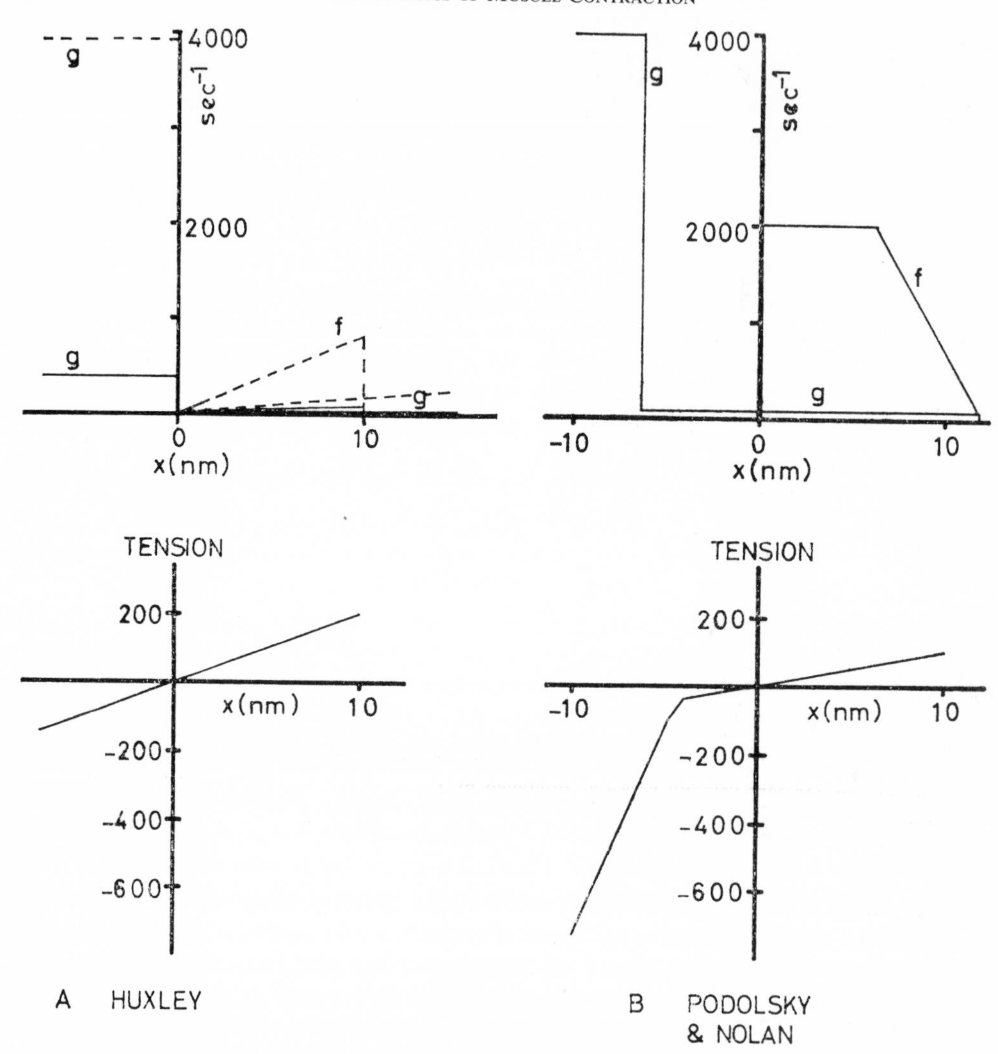

FIG. 30. Comparison of rate constants of attachment (f), detachment (g) and tension versus distortion (x) in (A) A. F. Huxley's (1957) formulation (absolute magnitudes from Julian, 1969) and (B) Podolsky's and Nolan's (1972) adaptation. The dashed lines in A are drawn magnified 10 times.

Here $n(x, t)$ is the fraction of actin sites having bridges attached with distortion x (defined above) at time t, $f(x)$ is the probability per unit time that a detached bridge with distortion x will attach at that distortion (note the special implicit definition here of distortion for detached bridges), and $g(x)$ is the probability per unit time that an attached bridge with distortion x will detach. The *total* time derivative of n, of course, includes the transfer of attached cross-bridges from one x value to the next at the rate of interfilament motion.

To account for contractile force, Huxley suggested that the force due to an attached bridge might be proportional to its distortion, and that attachment (controlled by $f(x)$) might occur preferentially at positive distortions. (Detached bridges could vary in their effective distortion due to thermal motion. It now appears unlikely (A. F. Huxley and Simmons, 1973)

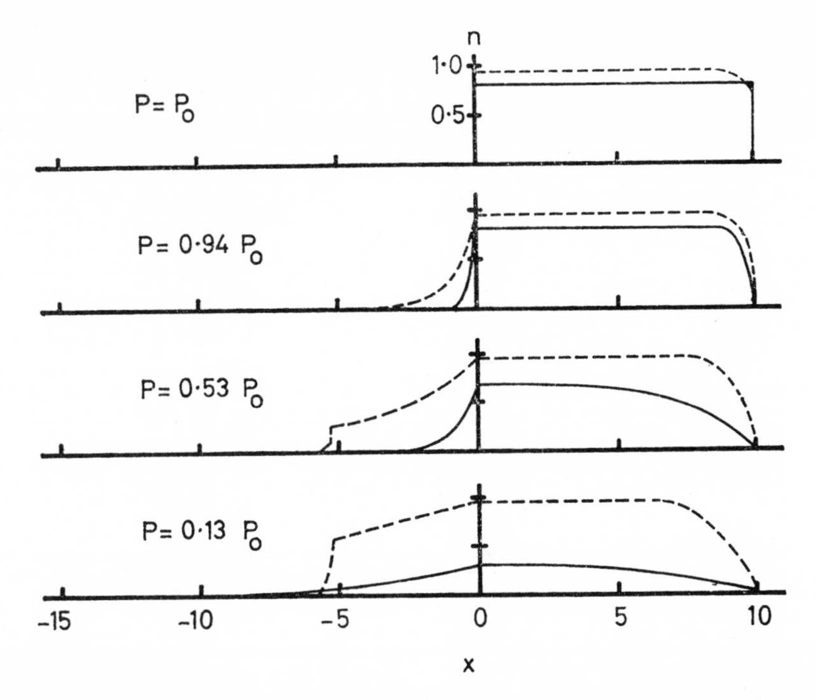

FIG. 31. Comparison of the equilibrium distribution of fraction of attached crossbridges n vs. distortion x at various steady tensions P (indicated as fractions of the full isometric tension P_0) during the constant-velocity phase as predicted by the A. F. Huxley (1957) model using the rate constant distributions shown in Fig. 30. Solid lines from A. F. Huxley (1957); dashed lines from Podolsky and Nolan (1972).

that this parsimonious assumption suffices in the light of the speed of contraction of certain muscles, but the 1957 formulation by no means stands or falls on this limb.) Further, efficiency dictates that the detachment rate $g(x)$ ought to be high when sliding has permitted the bridge to attain small distortion—where it exerts little force. The actual forms of the functions f and g which Huxley considered are shown in Fig. 30.

Confining himself to the constant-velocity case, Huxley obtained analytical expressions for the tension, shortening speed, and energy liberation. He showed that these conformed to A. V. Hill's (1938) measurements. The distributions of distortion for attached cross-bridges at several different shortening velocities are shown in Fig. 31. The higher the velocity, the greater the proportion of cross-bridges that have negative distortions, and which are resisting the shortening. The maximum shortening velocity is obtained when the contribution to the force from bridges with negative distortions just balances the force from those with positive distortions. Note that the theory predicts that as the shortening velocity increases, so the total number of bridges attached decreases.

We need not go into greater detail since A. F. Huxley and Simmons (1973) have reviewed very well the critical aspects of the 1957 scheme and their current view of it. Rather, we outline here the analyses of those who have adapted Huxley's formulation *in toto* and modified his rate-constant distributions to fit selected data.

These modifications will be more clear if we first describe qualitatively the way in which the formulation fits in with some of the dynamic relationships between length and tension.

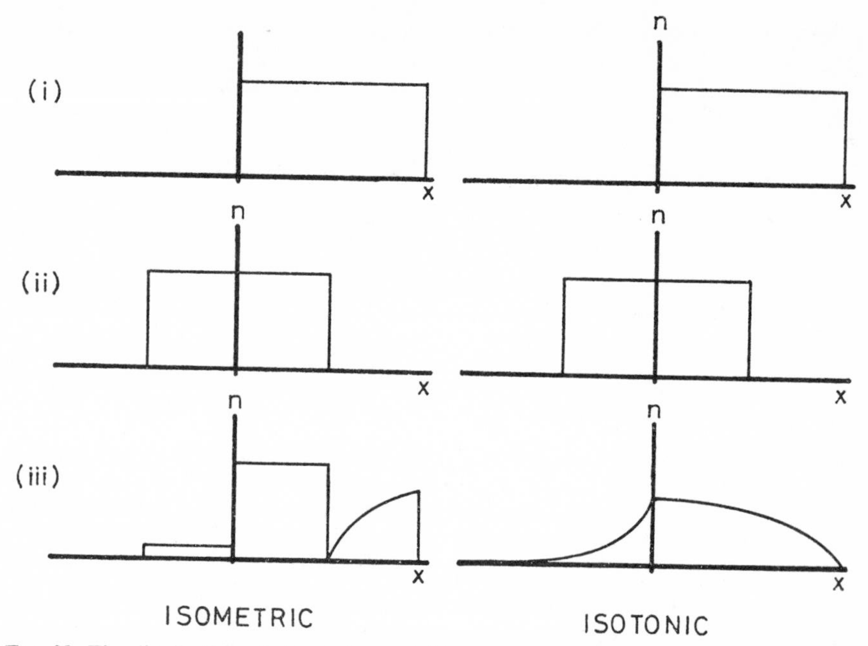

FIG. 32. The distribution of attached crossbridges n vs. distortion x before (i), immediately after (ii), and some time after (iii) application of a step length-change (isometric) or step tension-change (isotonic). The experiments correspond to the length-step and tension-step experiments of Fig. 4. The final equilibrium distribution in the isometric case is identical to the initial distribution, that in the isotonic case is as shown in (iii).

The effect of applying a step change in tension or length to the system is to shift the attached cross-bridges along the distortion axis by the amount necessary to change the tension to the new level for step changes in tension, or by the magnitude of the applied length change for a step change of length. If the tension-distortion curve for a cross-bridge is linear, then the instantaneous stiffness of the model is proportional to the number of attached bridges.

The new steady state, following the step change, will not be attained instantaneously since the distribution of cross-bridges along the distortion axis will not be that corresponding to the steady state. Consider the effect of applying a step length-change of that magnitude required to just reduce the tension in the muscle to zero. With Huxley's distributions this will require a length change of $h/2$, where h, as shown in Fig. 30, is the maximal value of the distortion for which the probability of attachment is non-zero. If the length of the muscle is now held constant and the filaments are rigid no sliding of filaments can take place. Cross-bridges will detach from the actin with the rate constant $g(x)$ relevant to their new distortion, and will attach as before in regions where $f(x)$ is non-zero. Those bridges that have negative distortions will detach rapidly, since g is large for negative x, resulting in an increase in tension (since they exert negative tensions). At any value of x the time course of change in fraction attached with time can be followed, and the time course of the overall tension change computed. The distribution of n will revert to its original distribution. The distributions before, immediately after, and some time after applying the step, are shown in Fig. 32.

A similar calculation can be made if the applied change is a step change in tension (see Fig. 32). In this case there will be length changes occurring at all times to counterbalance the change in distribution of attached bridges occurring because of attachment and detachment. The shape of the distribution of n can be followed as a function of time after applying the step, and will gradually approach a steady state as in Fig. 32.

Civan and Podolsky (1966) calculated the time course of the length changes following step changes of tension using Huxley's parameters for $f(x)$, $g(x)$, and bridge force. These are shown in Fig. 33 A. The steady-state velocity of shortening is not attained instantaneously, but is approached in a manner which, with Huxley's distributions, is not, as in the data, oscillatory.

Podolsky *et al.* (1969) and Podolsky and Nolan (1972, 1973) have therefore changed the dependencies of f, g, and tension upon distortion to ones which give rise to the oscillatory response. Their latest distributions are shown in Fig. 30, and the corresponding responses to step tension changes in Fig. 33 B. The oscillatoriness of the response is caused by the gap between zero distortion and the value of x at which the detachment probability becomes very high, aided by the steep increase in stiffness of attached cross-bridges in the region of negative x. The effect of the gap is to introduce a delay between the effect of bridge attachment (which occurs at a very high rate in regions of positive x) and bridge detachment. The steady-state distributions of n at different tensions are shown in Fig. 31. Note that, contrary to Huxley's original case, the number of attached bridges now *increases* as the velocity of shortening increases. In order to account for the response to step tension reductions, Podolsky and Nolan found it necessary to include a series elastic element, in series with their hypothetical cross-bridge, with stiffness approximately the same as that effected by the cross-bridges under isometric conditions. In other words, the instantaneous length changes resulting from a step change in tension from the isometric situation were about 50% accounted for by the cross-bridge elasticity, and about 50% by the series elasticity.

Note that the magnitudes of the probabilities in Podolsky and Nolan's distributions for f and g reach *very* much higher values than in Julian's (1969) estimates of Huxley's distributions. Thus in response to a step decrease in length there will be immediately a very rapid increase in tension due to the rapid bridge attachment in the region of positive x just vacated, plus a very rapid bridge detachment in the region of negative x where g is high if the step size was such as to shift the isometric distribution of attached bridges to this region. This will then be followed by a slower rise of tension whilst the bridges detach in the region of negative x for which g is low. At least qualitatively, the time course of tension recovery will follow that found by A. F. Huxley and Simmons (Fig. 26). Finally, it is important to note that Podolsky and Nolan (1972, 1973) do not treat at all the effect of extending the muscle.

Julian (1969) showed that A. F. Huxley's (1957) model can be adapted to account for the time course of tension changes in a tetanus and in a twitch, by assuming that the probability of attachment was proportional to the calcium-ion concentration, and using a plausible time-course of calcium release. He chose magnitudes of rate constants as shown in Fig. 30, and included a series elastic element to allow for the conditions of Jewell and Wilkie's (1958) experiments, which he used for comparison. Figure 34 shows another calculation of Julian's in which he has predicted the time course of tension development following a step decrease in length just sufficient to cause the tension to fall to zero, compared with Jewell and Wilkie's experimental results, and with the time course predicted from Hill's two-element model. The description of the data is obviously much more satisfactory than with Hill's model. As emphasized previously the velocity of shortening of the cross-bridge model is not a function of the load only, as in interpretations of Hill's model, but is dependent upon the past history of the muscle as described above.

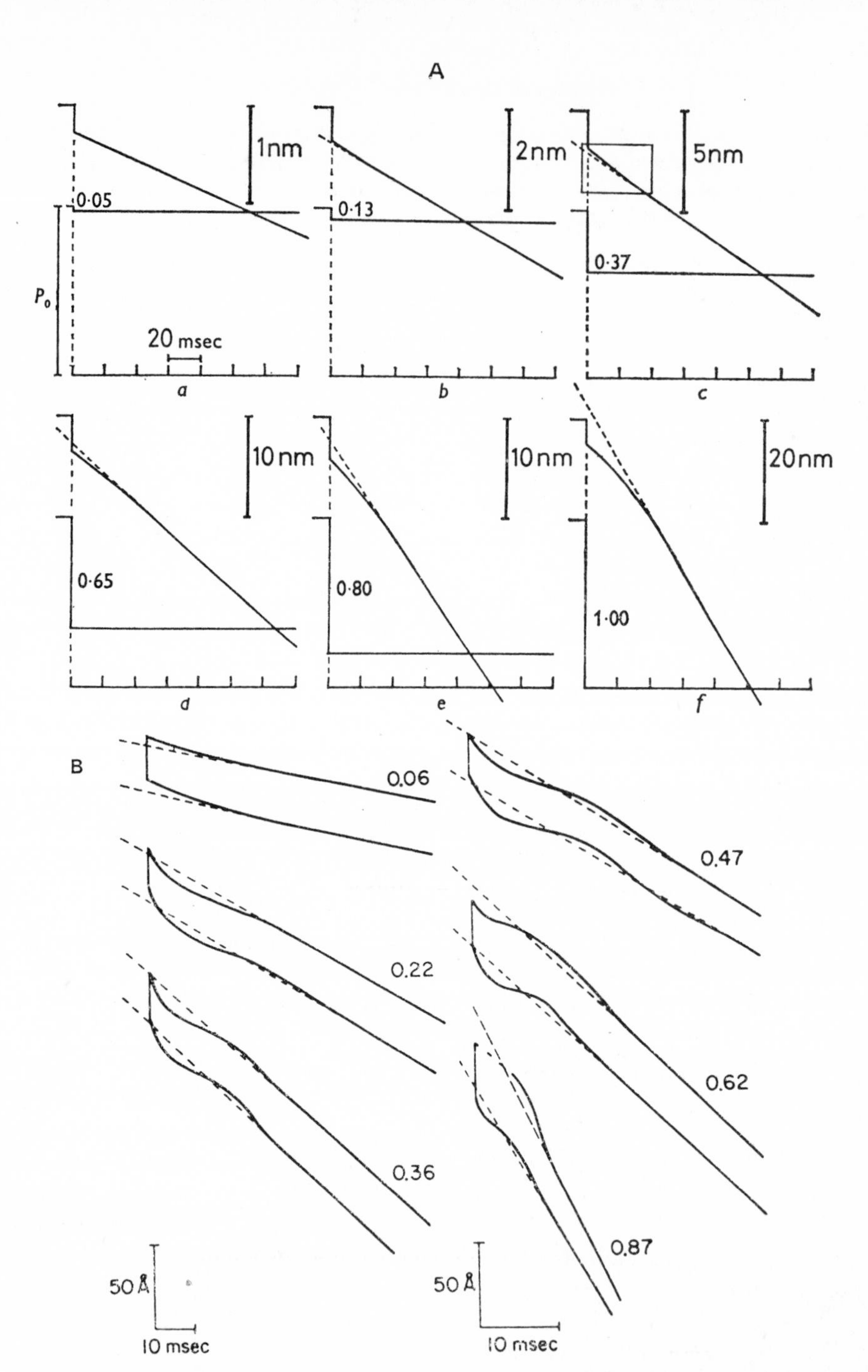

FIG. 33. Predictions of the time-course of the length changes obtained in a tension–step experiment on the Huxley (1957) model with (A), Huxley's (1957) distributions; (B), Podolsky's and Nolan's (1972) distributions. A is taken from Civan and Podolsky (1966) but with the scales modified to account for the magnitudes of the rate constants used in Fig. 30. In this figure the upper trace is of length and the lower trace of tension. In B, from Podolsky and Nolan (1972), the upper trace is the experimental length record and the lower trace is computed from the distributions of Fig. 30. Note that the initial "elastic" length changes have been omitted. On the model their magnitudes depend upon the amount of extra elasticity in series with the cross-bridge. In both parts the figures alongside the records denote the fraction of the full isometric force to which tension has been reduced.

In summary, both Podolsky and his coworkers, and Julian, have tested Huxley's original rate-constant distributions against non-constant-velocity experiments. Civan and Podolsky (1966) showed that Huxley's distributions were not capable of predicting the dynamics they observed in step-tension-reduction experiments, and Podolsky and Nolan (1972, 1973) found a new set of distributions that were capable of explaining these results. Julian (1969) tested Huxley's distributions against the results from the step length-change experiments of Jewell and Wilkie (1958) which were obtained from whole muscle and so provided a less sensitive test of the distributions than that used by Podolsky. Under these conditions, which required considerable extra series elasticity, Huxley's original distributions were fairly satisfactory. But Julian's distributions apparently cannot accommodate the earlier

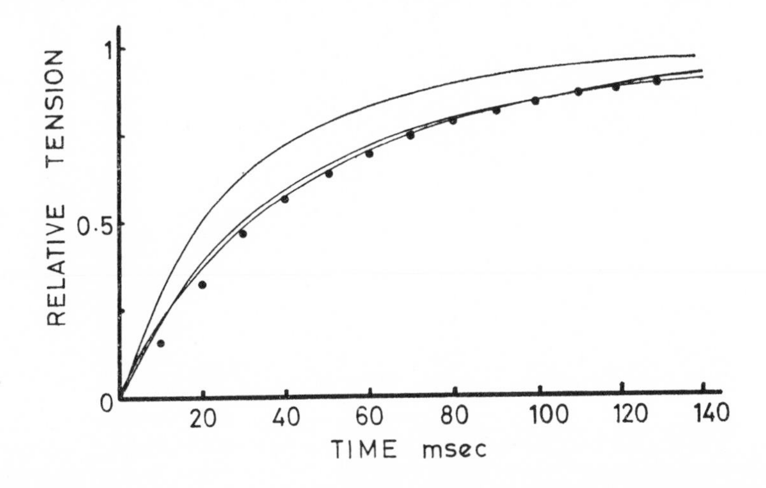

FIG. 34. Predicted time-course for the redevelopment of tension following a length step just sufficient to cause the tension to fall to zero during an isometric tetanus from Julian (1969), Descherevsky (1968), and on the Hill two-element model as computed by Jewell and Wilkie (1958) (top line) as discussed in § II. The circles are Jewell and Wilkie's experimental points.

data of Civan and Podolsky. Although no one has done the sums, it is probable that Podolsky's distributions would do equally well, if not better, in explaining Jewell and Wilkie's results. Recently, step-length-change experiments have been done by A. F. Huxley and Simmons (1971, 1973) with improved time resolution. Podolsky and Nolan's distributions give at least a qualitative description of these results, and could possibly be adjusted to fit quantitatively as well. With so many unknowns, it is a bit worrisome that most muscle dynamics seem mappable upon these rate-constant distributions, for we have no assurance at all that these explanations are correct in their specifics. Huxley and Simmons' recent scheme (below) offers a testable alternative to Huxley's original formulation of a more fundamental kind.

3. *The Formulation of A. F. Huxley and Simmons* (1971)

Section V detailed the experiments and their description upon which Huxley and Simmons have based their current cross-bridge hypothesis. Their results were illustrated in Figs. 25 and 26. If one assumes that the T_1 and T_2 curves are properties of the cross-bridges only, and that they do indeed reflect the properties of an elastic element in series with a damped

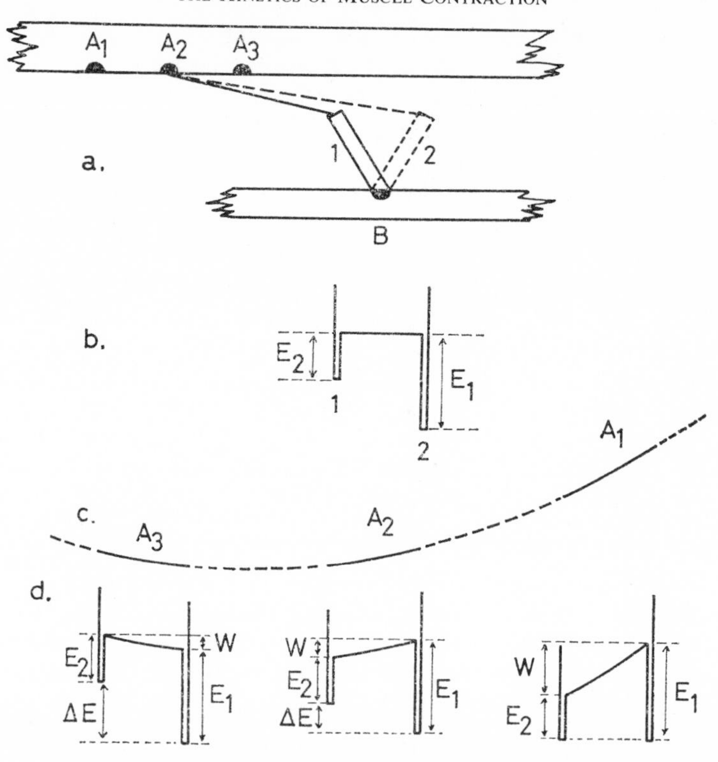

FIG. 35. A. F. Huxley and Simmons' (1971b) cross-bridge model, treated in terms of an elastic S_2 link in series with a two-state contractile S_1 head. a: The S_2 link stems from the A filament at A; the three spots A_1–A_3 represent different positions of A caused by sliding of the A filament relative to the I filament. The S_1 head can rotate about its attachment point on the I filament, B. b: Indicates the potential energy (p.e.) curve of the S_1 head as a function of its angle with respect to the I filament. c: Indicates the p.e. curve of the S_2 link *as a function of the angle of the S_1 head* for each of the three positions A_1–A_3. d: Shows the p.e. curve for the entire cross-bridge (i.e. the sum of parts b and c) for the three positions A_1–A_3. It has been assumed (for position A_3) that the link can exert compressive forces. The energy levels have been labelled as in A. F. Huxley and Simmons (1971b). The seemingly odd terminology (E_2 is the depth of the p.e. well for state 1) is made clear in their paper.

contractile element, then the tension–length curve of the contractile element is obtained as discussed in § V and illustrated in Fig. 27. A. F. Huxley and Simmons discuss qualitatively (1973) and quantitatively (1971b) an appealing mechanism producing this type of tension–length curve.

In order to illustrate their formulation, Huxley and Simmons used a cross-bridge model which contained a compliant S_2 link attached to a "contractile" S_1 head which could attach to the actin in a number of preferred configurations; just two of these are assumed in Figs. 35 and 37. We define "distortion" of the cross-bridge, as before, in terms of the relative displacement of A and B in Fig. 35. Thus for any given value of distortion there may be different configurations of the bridge depending upon the angle the S_1 head makes with the actin filaments. The potential energy (p.e.) vs. bridge angle for the head in the absence of the elastic link is as illustrated in Fig. 35 b. The elastic link has a p.e. curve vs. head angle as in Fig. 35 c. The p.e. curve of the combined element is taken as the sum of these two curves,

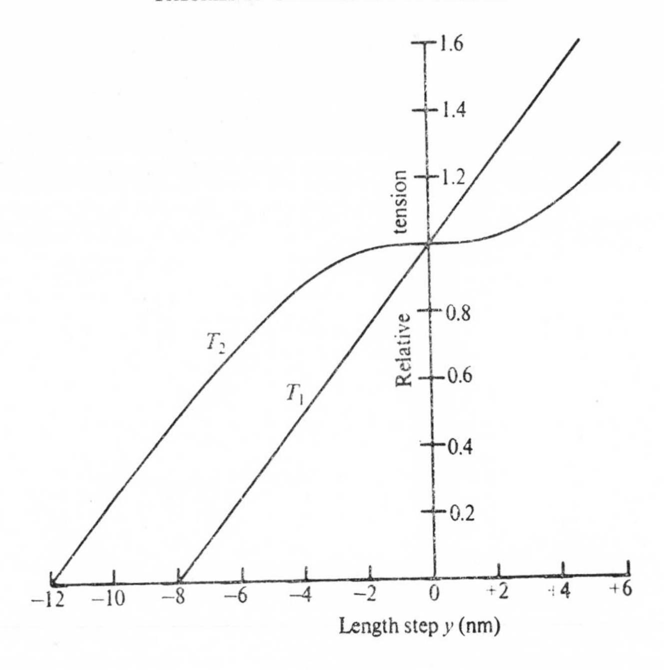

FIG. 36. Theoretical T_1 and T_2 curves (from responses to step length-changes) as produced by the model of A. F. Huxley and Simmons (1971b); the model is discussed in the text and illustrated in Fig. 35.

as in Fig. 35 d. Note that the total p.e. curve depends upon the extension of the elastic link due to the overall cross-bridge distortion. Three values of distortion are illustrated.

Under isometric conditions the attached bridges are assumed to be distributed among the various stable configurations according to a Boltzmann distribution.

If the A filament is moved rapidly with respect to the I filament the elastic element can change length before the contractile element can change configuration. The contribution to the total p.e. from the elastic link has thus changed, and the probabilities of transitions between the stable configurations of the contractile elements have likewise been changed. The equilibrium distribution of bridges amongst the various stable configurations will thus be different, and the bridges will redistribute themselves. The transitions will be governed by the rate constants between the various configurations which in turn depend upon the heights of the potential energy barriers between the states (the activation energies). In Fig. 35, with the elastic link at A_1, or A_2, this height is (E_2+W) for transitions from state 1 to state 2, and E_1 for the reverse reaction; with the elastic link at A_3 the corresponding heights are E_2 and $(E_1 + W)$. The steeper the p.e. curve of the elastic link the slower the reaction between states 1 and 2, and thus the slower the time course of the tension change following a step change of length of the elastic link to this position on its curve. The potential energy curve of the elastic link will be steeper the greater the tension in it. It follows that the change of tension following a step increase of length will be slower than that following a step decrease, as observed experimentally.

A. F. Huxley and Simmons (1971b) treat the simplest case of two states, and demonstrate that the shape of the T_1 and T_2 curves produced by the model (Fig. 36) have the same general form as those obtained experimentally.

By the mechanism described above, length changes producing interfilament motion can affect the transitions between different states of the contractile element. It is this suggested interaction between an experimentally controlled variable and the operation of the contractile mechanism that makes Huxley and Simmons' formulation inherently testable.

We find it illuminating to describe this system in another way. For each of the two states of the S_1 head upon the I filament one can draw the potential energy of the cross-bridge as a function of the distortion. For the case in which the angle of the head does not change whilst the bridge remains in one state, the resulting curve will be just that of the elastic link, as illustrated in Fig. 37, curve 1. A similar curve for the second state of the S_1 head will be displaced along the y axis by an amount equal to the separation, in the direction of the

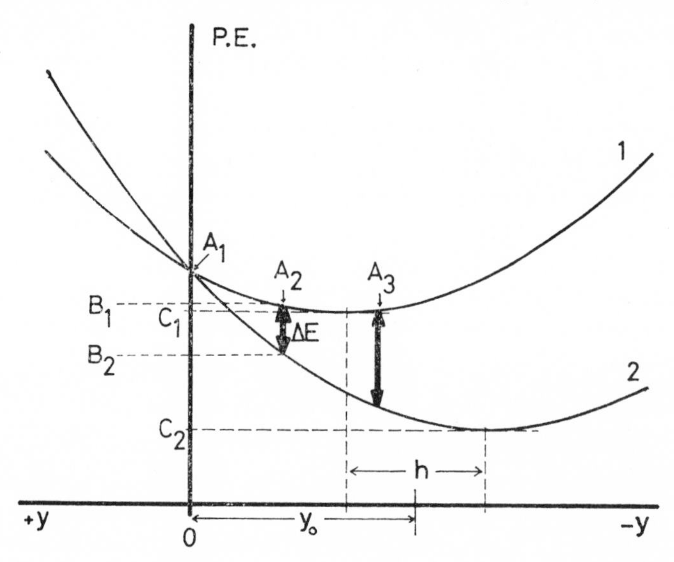

FIG. 37. Predicted potential energy curves of the elastic component of the cross-bridge vs. its linear extension for two positions of the contractile component. The y axis thus corresponds approximately to bridge distortion. The lettering is as discussed in the text.

filaments, of the two states of the S_1. The curve may also be displaced vertically. A. F. Huxley and Simmons (1971b) assumed that when the mean extension of the elastic link took on a particular value y_0, the energies of the two states were identical, and thus the states were equally populated. The p.e. for state 2 is then as shown in curve 2 in Fig. 37. y_0 by implication is the separation between the position of the head at the instant of attachment and the mean position of the head between the two attached states being discussed. A delta function is implied in their formulation for the probability of attachment versus cross-bridge distortion.

For either of these descriptions one can determine the equilibrium distribution of attached bridges between the two states for any value of distortion (y) of the cross-bridge, as well as the dependence on y of the rate constant for return to equilibrium of this distribution if it is disturbed. A. F. Huxley and Simmons (1971b) give a quantitative treatment based on Fig. 35. With Hookean springs of stiffness K, and the formulation of Fig. 37, the energy levels for position A_2 are:

$$B_1 = \tfrac{1}{2}K(y + y_0 - h/2)^2 + C_1 \qquad (11a)$$

$$B_2 = \tfrac{1}{2}K(y + y_0 + h/2)^2 + C_2. \tag{11b}$$

When $y = 0$, $B_1 = B_2$. Substitution into the above equations gives

$$C_2 - C_1 = Khy_0. \tag{12a}$$

The general case gives

$$B_2 - B_1 = Khy. \tag{12b}$$

At equilibrium the distribution between the two states is given by the Boltzman distribution. If n_1 and n_2 are the fractions of attached bridges in states 1 and 2 (thus $n_1 + n_2 = 1$), and k_+ and k_- are the rate constants for the transitions between states (1 and 2) and (2 and 1), respectively, then

$$n_2/n_1 = k_+/k_- = \exp(-(B_2 - B_1)/kT) = \exp(-Khy/kT). \tag{13}$$

This is identical to eqn. (8) of Huxley and Simmons, and their further development is now applicable.

As A. F. Huxley and Simmons (1973) are at pains to point out, the assignment of the elastic link and contractile element to the S_2 and S_1 subfragments, respectively, is not the only one possible. They list several alternatives.

It is instructive to formulate these carefully since such alternatives may be distinguishable via X-ray diffraction. For example, consider the following variant of one of Huxley and Simmons' alternatives: S_2 is assumed rigid, and the S_1 head, as before, can exist in one of two states. The minimum-potential-energy positions of the two states are different, and for each state the head exerts a restoring torque proportional to the angular displacement from its minimum-potential-energy position. This latter property replaces the elasticity of the S_2 link in the first formulation.

If suitable assumptions are made about the activation energies, the calculation of the theoretical T_1 and T_2 curves proceeds much as it does for Huxley and Simmons' case. Here, however, note that a change of state need not be accompanied by a change in the overall configuration of the S_1 head with respect to the actin. Rather it may be a more microscopic "internal" change. But since the S_2 is rigid, the rapid "elastic" step changes here (by definition) do involve a change of orientation of the S_1 head, so that this alternative may be distinguishable from the elastic-S_2 case by X-ray diffraction measurements. Figure 38 shows the way in which angle changes of the S_1 head upon the I filament occur for the two alternatives discussed.

The explanation by Huxley and Simmons of the initial very rapid events occurring following step length or tension changes, in terms of transitions between discrete states of an attached cross-bridge, differs fundamentally from that of Podolsky and Nolan (1972, 1973), who assume very rapid attachment and detachment to occur following applied step changes. Not only are the explanations alternatives, but there is a fairly straightforward test to distinguish them—that is, to measure in some way (e.g. with a second step) the number of attached bridges at each instant following a step length or tension change.

Pitts, Nolan, and Podolsky (quoted in Podolsky and Nolan, 1973) present records from a double-step experiment in which step changes in tension are applied. The first step is applied from the full isometric tension and the second step is applied about 40 msec later, during the period of steady shortening in the fibres. Analysis of the initial length changes (those in phase with the applied tension changes) are difficult, since the time resolution of the apparatus does not permit clear measurements at these short times; however, by two criteria—(1) extrapolating the length records back to the time of application of the step,

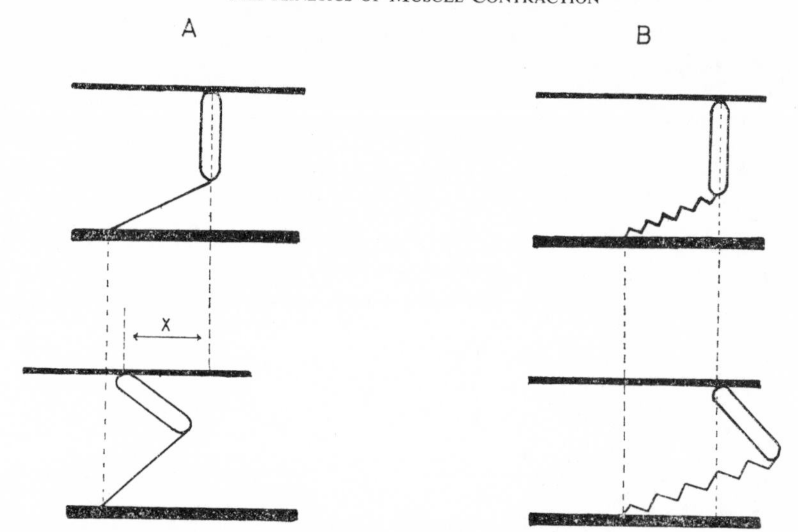

FIG. 38. Ways in which changes of angle of the S_1 head relative to the I filament could occur when (A) the elastic and contractile contributions are both within the S_1 head and (B) the elastic element is within the S_2 link and the contractile element is within the S_1 head. In (A) a relative sliding of the filaments is required; in (B) the change in angle occurs when the contractile head changes state.

(2) extrapolating the constant-velocity phase back to the same time—a series of three estimates can be obtained from their fig. 8 (parts a, b, and c) for the "instantaneous" stiffness of the fibres in the isometric state and two obtained (parts a and b) for the stiffness during steady shortening. By either criterion the stiffness is less when the step is applied during the steady shortening. On the other hand, their model predictions (in their fig. 7) when analysed by either of the above two criteria give a greater stiffness when the step is applied during the steady shortening than when applied during the isometric phase. Their model gives good fits to the records after the initial step.

A. F. Huxley and Simmons (1973) also present an experiment suggesting that during the period of steady shortening there is a smaller number of bridges attached than in the isometric muscle before the tension change. However, as Podolsky and Nolan point out (personal communication), the evidence is not conclusive from this record, since the response to the step increase in tension, used to sample bridge number, might be qualitatively different from those to step decreases.

Thus the available evidence, which is not conclusive, suggests that during the period of steady shortening there are less bridges attached than during the isometric state, in contradiction to the prediction of Podolsky and Nolan's adaptation of Huxley's (1957) model.

The model of A. F. Huxley and Simmons (1971b, 1973) describes only the form of the response in the first few milliseconds after application of a step change in length. A full description must obviously treat the entire response, and involve bridge attachment and detachment. Julian *et al.* (1973) have extended the model of Huxley and Simmons to account for such detachments and attachments, making plausible assumptions about the dependence of the rate constants of attachment and detachment upon bridge distortion, and have demonstrated that the general form of the observed experimental response can be reproduced.

4. *Can Filament Elasticity be Ignored in Vertebrate Striated Muscle?*

Both A and I filaments have been assumed rigid in all analyses of models of vertebrate muscle. Thus all cross-bridges along the sarcomere have been lumped into a single ensemble without treatment of possible systematic variation of such parameters as distortion along the overlap region.

The reason this simplifying assumption has not been made for insect flight muscle is that local filament strain itself has been, at least plausibly, a critical variable in the process of strain activation. Methods have therefore been developed to study these distributed effects (Thorson and White, 1969).

In order to evaluate the effect of compliance in the filaments quantitatively we require estimates of (1) the stiffness of the A and I filaments, (2) the bridge force under isometric conditions when the distortion is zero, and (3) the cross-bridge stiffness during distortion. It is convenient to express the filament stiffnesses as stiffness times unit length, and the cross-bridge force and stiffness as though the cross-bridges were distributed uniformly along the length of the A filament, in terms of bridge-force and bridge stiffness per unit length. White (1967) obtained estimates of these parameters for insect flight muscle from a variety of structural and mechanical data; these values are shown in Table 3.

TABLE 3. VARIOUS MECHANICAL PARAMETERS

	From White (1967) Insect flight muscle	From text Vertebrate muscle	Units
Cross-bridge force (ϕ_0)	600	450	pN/μm
Cross-bridge stiffness (q)	8×10^4	4×10^4	pN/μm^2
A filament stiffness times unit length (k) (lower bound)	1.5×10^4	3.5×10^4	pN
I filament stiffness times unit length (λ) (lower bound)	2×10^4	1.8×10^4	pN

The above estimates are approximate, and derived independently for the two cases; no significance should be attached to differences between the insect and vertebrate values.

One can also obtain estimates for vertebrate muscle from available data. Maximum isometric tensions in vertebrate muscle are about 2 kg/cm^2 (= 0.2 N/mm^2). An A-filament separation of 45 nm (Elliott *et al.*, 1963) gives approximately 6×10^8 A filaments per mm^2 (assuming that the entire cross-section is occupied by filaments), and thus the cross-bridge force is about 340 pN per A filament. In one half-sarcomere the cross-bridge region of the A filament is about 0.75 μm long, giving a value for bridge force per μm of A filament (ϕ_0) of 450 pN/μm. If the bridge force is reduced to zero by a cross-bridge distortion (on release) of 12 nm, then, assuming for the moment that the stiffness is Hookean, the stiffness per unit length (q) is 4×10^4 pN/(μm)^2. Lower bounds can be given to the filament stiffnesses from the limits of accuracy of the X-ray diffraction measurements, since these fail to detect changes in the filament spacings with the tension borne by the filaments. We have assumed that the evidence does not exclude a change in the mean strain of either the A or the I filaments of 0.5%, comparing the spacing under full isometric tension with that under low tension.

From eqn. (14) below, we estimate the mean strain in the A filament to be $\phi_0 b/2k$, where ϕ_0 is the bridge force per unit length, b is the half-sarcomere length, and k is the A filament stiffness times unit length. The above values yield $k = 3.5 \times 10^4$ pN. The same argument, of course, leads to half this value for the lower limit of the I filament stiffness times unit length. These estimates are also given in Table 3. They differ very little from those for insect flight muscle. The value of q is also compatible with that derived from A. F. Huxley and Simmons' (1971b) estimate of the S_2-link stiffness in their formulation.

Although the average filament strains implied by these estimates are small, it is not at all clear that the associated variations of distortion along the overlap zone will be negligible. To see this worry, note (as in Fig. 39) that both tension and strain at the *free* ends of the A and I filaments are zero, whereas the full tension and strain appear at their ends of connection at the Z and M lines.

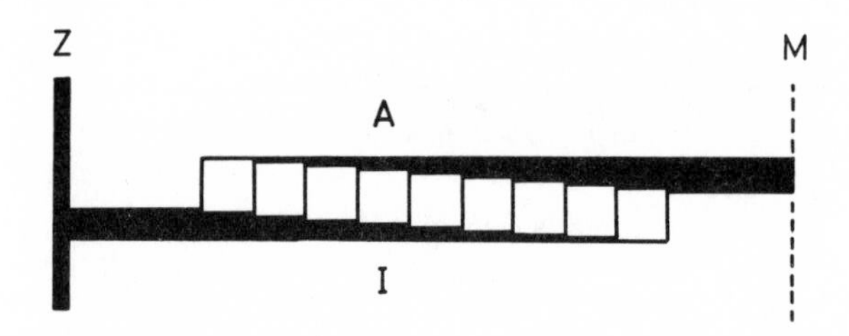

FIG. 39. Distribution of tension and strain along the A and I filaments when interfilament shear force is produced by cross-bridges. The magnitude of local filament tension and strain are indicated by the width of the lines representing the filaments.

The obvious control is to repeat some of the vertebrate model calculations with the distributed methods of Thorson and White (1969). A sufficiently simple case had not presented itself, however, until the quick-step experiments and novel interpretation of A. F. Huxley and Simmons (1971a, b), discussed above, appeared. In the following, we show that distortion can be distributed sufficiently to require analysis, and that the type of non-exponential relaxation of tension between T_1 and T_2, observed in many of Huxley and Simmons' experiments, is thereby predicted on the basis of their hypothesis.

Assume full overlap of the filaments, and (as in Thorson and White, 1969) that $y(x)$ and $z(x)$ are the displacements from zero strain of points on the A and I filaments respectively, where x is distance along the overlap region. The A filament is free at $x = 0$ and the I filament free at $x = b$. $\phi(x)$ is the local interfilament shear force per unit length; k and λ are the stiffnesses times unit length of the A and I filaments.

Consider first the case for which $\lambda \gg k$ so that only A filament strain is appreciable and $z(x) = 0$; at the activated equilibrium prior to a step change of length, $\phi(x) = \phi_0$ (the undistorted attached bridge force per unit length) since the filaments are stationary and all distortions have "decayed" to zero in the process of attachments and reattachments. The corresponding steady displacements $y(x)$ are readily shown to be

$$y(x) = \frac{\phi_0}{2k}(x^2 - b^2) + L, \tag{14}$$

where L is the applied extension, i.e. $y(b) = L$.

If now a length step ΔL is applied, and the arrangement is studied before attachments or detachments (or changes of state in Huxley and Simmons' scheme) can occur, the resulting tension change ΔT is due to the extra shear forces due to the various elastic distortions (or S_2-link elongations) of bridges, $\Delta\phi(x)$; $\Delta\phi(x) = q\Delta y(x)$, where q is the cross-bridge stiffness per unit length, assumed to be constant in these calculations. The variation of these distortions $\Delta y(x)$ with x is calculated from the differential equation for the physical situation following the step,

$$\frac{d^2\Delta y(x)}{dx^2} - \frac{q}{k}\Delta y(x) = 0, \tag{15}$$

which, it would be remiss not to point out, is actually a degenerate form of Hill's differential equation; see, for example, Murphy (1960). Since $\Delta y(b) = \Delta L$ and $d\Delta y(0)/dx = 0$, the solution is

$$\Delta y(x) = \Delta L \frac{\cosh\sqrt{q/k}\,x}{\cosh\sqrt{q/k}\,b}. \tag{16}$$

As $k \to \infty$, the $\Delta y(x)$ all approach ΔL as they ought to do.

The width of the Δy distribution will, of course, be reduced somewhat if the filament stiffnesses k and λ are of comparable size. Here one obtains a third-order differential equation after eliminating $\Delta z(x)$ (Thorson and White, 1969):

$$\frac{d^3\Delta y}{dx^3} - q\left(\frac{1}{k} + \frac{1}{\lambda}\right)\frac{d\Delta y}{dx} + \frac{q\Delta T}{k\lambda} = 0. \tag{17}$$

The above boundary conditions plus the two additional conditions that

$$\frac{d\Delta y(b)}{dx} = \frac{\Delta T}{k} \quad \text{and} \quad \frac{d^2\Delta y(0)}{dx^2} = \frac{q}{k}\Delta y(0)$$

serve to eliminate the three constants of integration and ΔT, so that the distributed distortion $\delta(x)$ immediately following a step ΔL from equilibrium is

$$\delta(x) = \Delta y(x) - \Delta z(x) = \frac{\Delta L(\lambda + k)\,[(1/k)\cosh(\alpha x) + (1/\lambda)\cosh(\alpha(x-b))]}{2 + [(\lambda/k) + (k/\lambda)]\cosh(\alpha b) + (\alpha b)\sinh(\alpha b)}, \tag{18}$$

where

$$\alpha = \sqrt{\left[q\left(\frac{1}{k} + \frac{1}{\lambda}\right)\right]}. \tag{19}$$

As $\lambda/k \to \infty$, this function reduces appropriately to the $\Delta y(x)$ distribution given above.

The departure of this distribution of distortion from uniformity is illustrated in Fig. 40 for plausible ranges of values of λ and k. A further test of the notion that these effects are operative is to ask what effect the distributed distortion has upon the calculated transient recovery from T_1 to T_2 in Huxley and Simmons' scheme. The full dynamics of the recovery are complex, and the above calculation does not take into account that T_0 and T_2 differ somewhat. However, a first estimate of the non-exponential relaxation of the distortion-induced tension $\Delta T(t)$ is available if one assumes simply that each segment dx of bridges at position x contributes to $\Delta T(t)$ an exponentially decaying shear force of magnitude proportional to $\delta(x)$ and with a rate constant given by the initial distortion $\delta(x)$ at that x.

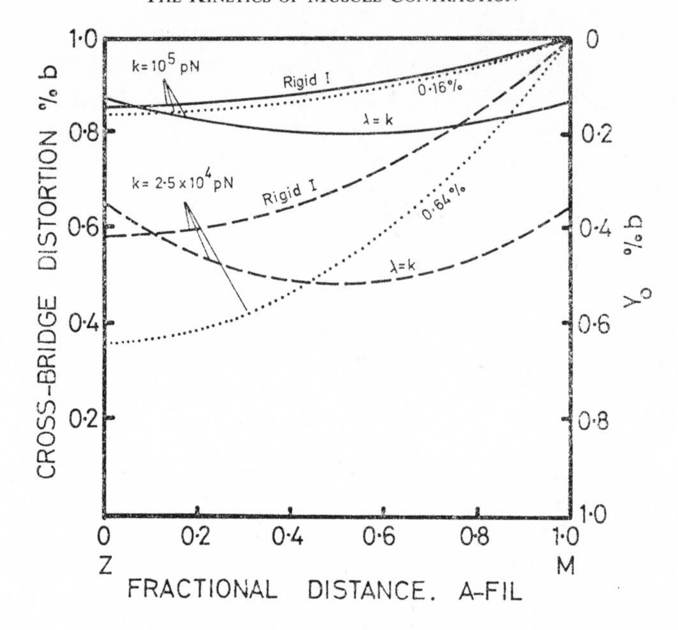

FIG. 40. Distribution of cross-bridge distortion as a function of position along the ½A filament under the conditions described in the text, for a length change of 1% of the half-sarcomere length (b). The curves labelled "Rigid I" have been calculated assuming that the I filament stiffness is infinite; those labelled "$\lambda = k$" have equal A and I filament stiffnesses. The dotted lines indicate the displacement y_0 from zero strain along the A filament before the application of the length change. The numbers against these curves are the values of the mean strain in the A filament under conditions of full isometric tension for the two values of k used here.

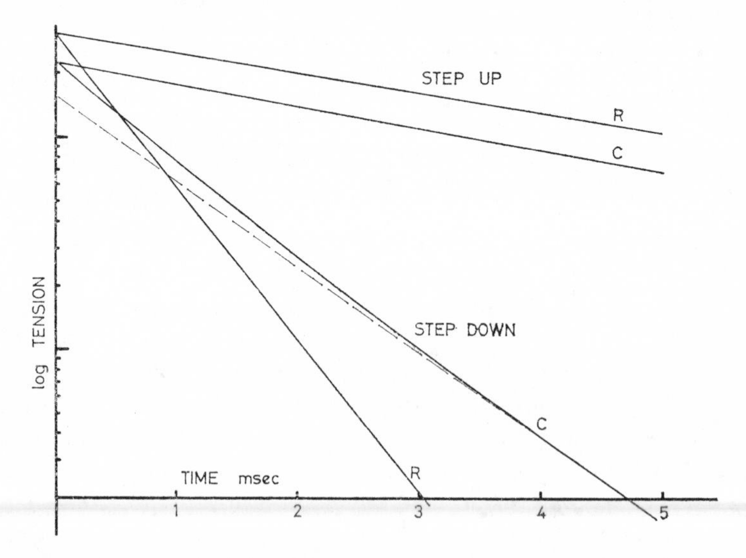

FIG. 41. Effect of filament elasticity on the time-course of the tension changes following a step change of length, as described in the text. R denotes the response with both filaments rigid; C the response with A filaments compliant (stiffness 2.5×10^4 pN) and I filaments rigid. The dashed line is the asymptote of the step-down transient; the transient is non-exponential because of the distributed distortion.

Since in Huxley and Simmons' scheme this rate constant depends upon δ (y in the terminology of Huxley and Simmons) as

$$r(\delta) = a(1 + e^{-c\delta}), \tag{20}$$

where a and c are constants, $\Delta T(t)$ is no longer a simple exponential $e^{-r(\Delta L)t}$ but rather an integral over the differing relaxations along the overlap region:

$$\Delta T(t) = K \int_0^b \delta(x)\, e^{-r[\delta(x)]t}\, dx, \tag{21}$$

where K is constant.

Numerical solutions for $\Delta T(t)$ are plotted in Fig. 41. As one might have intuited, for releases, the small distortions in the distribution contribute early, more rapid, components to the decay, which appears more truly exponential as these settle out. The reduced dependence of rate constant on stretch makes the effect less obvious for extensions. Although the comparison is preliminary and awaits measurements with improved time resolution, such an effect is discernible in the measurements of A. F. Huxley and Simmons (1971b), as Abbott (1972; see also A. F. Huxley and Simmons, 1972) has discussed in a different context. It should also be clear that the effect of filament elasticity is not the only explanation available for analogous distributed effects. For example (A. F. Huxley, personal communication), one might have distributed distortion at constant length due to mismatch of the bridge and actin-site periodicities, even with sensibly rigid filaments.

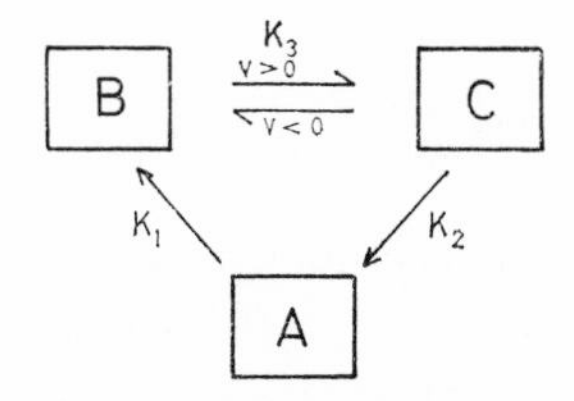

FIG. 42. Descherevsky's three-state model. The transitions between states B and C are caused by filament sliding, not by biochemical transitions, and are dependent upon the velocity of relative sliding.

5. *Descherevsky's Formulation*

Descherevsky (1968) has produced a scheme for cross-bridge activity based upon a treatment keeping count of the fraction of the cross-bridges attached. His paper gives a very clear account of the problems inherent in such a treatment.

Descherevsky made certain simplifications which lead to a three-state model, with two attached states and one detached state. Rather than treat distortion by the methods used by Huxley, Podolsky and ourselves (which require numerical computation for all but the simplest situations), Descherevsky made the transitions between his two attached states dependent only upon the velocity of sliding of the filaments. The cycle is shown in Fig. 42. A cross-bridge in state B generates + 1 unit of force, and in state C generates − 1 unit of force.

Descherevsky's treatment of the effects of cross-bridge distortion purely in terms of a transition between two attached states, which depends only on cross-bridge movement, results in a physically unattractive model. For example, under isometric conditions there can be no cycling of cross-bridges and hence, on most assumptions, no ATP hydrolysis. Nonetheless, it is procedurally interesting that the steady-state force-velocity relationship yields Hill's formula; the response to a step decrease in length (shown in Fig. 34) results in a curve similar to that obtained by Julian (Descherevsky has a series elastic element similar to Julian's). The distinction between sufficiency and necessity could not be better illustrated.

6. *T. L. Hill's Analysis*

T. L. Hill (1968a, b) and T. L. Hill and White (1968a, b) have considered the statistical mechanics of the sliding-filament and cross-bridge theories. Paper II in this series of four (T. L. Hill, 1968b) is the most general and ought most to influence future cross-bridge formulations. Hill has shown (see also T. L. Hill, 1966) how one can formulate and diagram such models so as to apply the fundamental constraints which exist in principle between the rate constants (e.g. $f(x)$ and $g(x)$) and the bridge forces (also functions of x). These are interrelated via the relevant partition functions vs. x. The problem at the moment (T. L. Hill, personal communication) is that the candidate cycles are not sufficiently specified experimentally; the x-dependence of unknown small "backward" rate constants may "soak up" the constraint one would have if they were known (but it is worth while checking whether they in fact could). In any event, these constraints ought to become critical as the facts of the cycle come to the surface (see biochemical correlates, § IV).

VII. INSECT FIBRILLAR FLIGHT MUSCLE

1. *Introduction*

Pringle (1967) and Tregear (1973) give excellent introductory accounts of the properties of intact insect fibrillar flight muscle.

In the living insect the fibrillar (synonymously, myogenic or asynchronous) flight muscle is characterized by its ability to drive the wings in the flight cycle at a much higher frequency than that of the nervous input to the muscles (in *Calliphora*, Pringle (1949) recorded a wing-beat frequency of 120 Hz, whereas the nervous input is about 3 Hz). The frequency of mechanical output from the muscles is determined by the natural frequency of oscillation of the mechanical system driven by the muscles (the effective inertia of the wings and the elasticity of the muscles and thorax of the insect). The muscle is able to drive this system because when active its response to length-changes, at the frequency of wing-beat in the intact animal, is to produce delayed tension-changes. The muscle thus acts as a "negative viscosity", enabling the oscillations to be maintained. The term "oscillatory work" is often used to denote the work produced when the length is oscillated sinusoidally.

Machin and Pringle (1960) determined the mechanical properties of the isolated live basalar muscle fibres from the beetle *Oryctes*, applying a sinusoidal length input of varying frequency and measuring the resulting tension changes. They demonstrated that tension was delayed behind the applied length changes over a range of frequencies (about 5–60 Hz), producing maximal power at about 30–40 Hz.

Jewell and Rüegg (1966) demonstrated that glycerol-extracted fibres from the water-bug *Lethocerus* exhibited similar properties when immersed in suitable solutions, and thus that the ability of the muscle to produce oscillatory work was a property of the contractile proteins and was not due to the normal membrane control systems of the fibre, which in any case are very sparse in fibrillar muscle (Smith, 1966). (It is the lack of sarcoplasmic reticulum in such muscles that enables the myofibrils to be separated so easily, thus giving the muscle its fibrillar appearance.)

For a variety of reasons nearly all the work on *Lethocerus* has used the glycerol-extracted preparation. There are a number of advantages in using such a preparation. The normal nervous control system is abolished because the membrane systems are disrupted, and thus the chemical environment of the fibres can be controlled in a way not possible with live muscle. Measurements of ATPase activity are made very much simpler, the effect of changing concentrations of any desired chemical can easily be determined, and many experiments can be done with the muscles of a single animal (not a trivial factor with tropical animals available only during restricted seasons of the year).

Although no detailed comparison of the live and glycerol-extracted preparations has been made, Thorson and White (in prep.) have shown that freshly dissected *Lethocerus* fibres, placed in the solutions and apparatus used for glycerol-extracted fibres, can exhibit activated frequency-response curves (in the range 1–40 Hz) nearly identical to those for glycerol-extracted muscle—both before and after the single fibres are pared down to bundles of myofibrils. The fresh fibres seem hardier, easier to pare down without loss of work production, and may well differ in other crucial respects. This point should be borne in mind when comparing the responses obtained from insect flight muscle with those described in the previous sections on vertebrate striated muscles, which have all been measured for live muscle.

2. *Experimentally Determined Properties of Glycerol-extracted Insect Fibrillar Flight Muscle*

(i) In the resting state, insect flight muscle is very stiff by comparison with vertebrate striated muscle (Machin and Pringle, 1959). This property is maintained in the glycerol-extracted state (White, 1967). The "short-range elastic component" of D. K. Hill (1968), which Hill ascribes to cross-bridge interaction in vertebrate muscle, has an elastic limit at less than $\frac{1}{2}$% strain in normal Ringer's solution. This is much less than the 9% elastic limit of insect muscle found by White (1967). There is no change in the low ATPase activity of the relaxed muscle with applied strain.

The only structure described that could account for the high stiffness of the insect fibrillar flight muscle is the C filament (discussed in § III) between the *Z* line and the end of the A filament. However, the finding of paramyosin in the insect flight muscle (Bullard *et al.*, 1972) in amounts of about 6–7% of the myosin present in *Lethocerus*, has led to speculation that whatever is giving rise to "catch" in such muscles as the anterior byssus retractor muscle of *Mytilus* (which contains large amounts of paramyosin) might also be related to the maintained stiffness of the resting insect muscle. Against this idea is the finding that the resting stiffness of the insect fibres is obtained up to strains of 9% (White, 1967), which is too great a value for maintained cross-bridge interactions.

(ii) In the activated state small applied length changes from zero up to 3–4% above rest length (defined as that length at which the tension in the relaxed fibres is just zero) result both in large changes in maintained tension (Jewell and Rüegg, 1966; White and Thorson, 1972) and in maintained ATPase activity (Rüegg and Tregear, 1966; Rüegg and Stumpf, 1969a). The stiffness of the active fibres for the maintained changes can be about three times that for the relaxed fibres (White and Thorson, 1972). The ATPase activity of the fibres can be increased by a factor of about 3 for a 3–4% increase in length. These changes cannot be accounted for in terms of changes of the extent of overlap of the A and I filaments, which at rest length is with complete overlap of the cross bridges by the I filament. (For "maintained" above, one should read "maintained for many minutes" as there is slow stress relaxation in glycerol-extracted fibres; White, 1967.)

These maintained effects are not observed in vertebrate striated muscle, which would show much smaller changes in either maintained tension or maintained ATPase activity under the conditions of the experiment.

At rest length the ATPase activity at saturating levels of Ca^{++} is much less than the activity obtainable at greater lengths. In other words, the full activity of the intact contractile system cannot be induced by calcium ions alone, as seems to be the case with vertebrate striated muscle.

We will call this ability of the insect flight muscle to produce maintained changes of activity with maintained length changes "strain activation".†

The mechanism of strain activation is not known. Its maintenance, at least over many tens of minutes, implicates some structural continuity across the sarcomere; any familiar form of cross-bridge interaction, postulated to mediate the effect, which is compatible with the high ATPase observed would not maintain the strain activation because of recycling of attachments. Thus it is tempting to suggest that C filaments connected to the A filament permit the change in activation. The paramyosin could in that event be an intermediary, with stress on the paramyosin modifying a paramyosin–myosin interaction. In this connection it is of interest that paramyosin inhibits the ATPase activity of actomyosin made from dung-beetle muscle myosin and rabbit actin (Hammond and Bullard, personal communication). Szent-Gyorgyi *et al.* (1971) made similar observations on molluscan catch muscle. A second candidate explanation, for which the necessary structure has not been described, is that a change of length of the sarcomere causes a modification of the position of the tropomyosin, thereby enabling a change in activity. Tropomyosin, in the absence of troponin, strongly activates the ATPase activity of insect flight muscle myosin mixed with either insect or rabbit actin at low ionic strength, over the range of calcium-ion concentrations 10^{-9}–10^{-5} M (Bullard *et al.*, 1972). Some form of continuity of the I filament, not staining under the electron microscope, would then seem to be required.

(iii) If the length change discussed in the previous section is applied very rapidly (in about 1 msec or less), then characteristic transients precede the final tension. The nature of these transients depends on the concentration of inorganic phosphate (P_i) in the solutions bathing the contractile proteins (White and Thorson, 1972):

(a) In the presence of at least 5 mM free phosphate, there is an initial rapid transient rise in tension, which decays with a time constant of a few msec, followed by a delayed rise in tension having a roughly exponential time course with a time constant of about 30 msec (Fig. 43 B).

(b) In the absence of P_i the response differs from that in (a) particularly for step heights greater than a few tenths of a per cent in amplitude (Fig. 43 A). There is an initial rapid transient response, followed by a delayed tension change, but this delayed

† Actually, as a strained fibre "stress relaxes" (tension falls off slowly over tens of minutes), the ATPase declines with tension even though *fibre* strain is maintained (Rüegg and Stumpf, 1969a, fig. 3). We do not know what structure is stress relaxing over long periods; we have therefore not adopted the term "stress activation" but rather retained "strain activation" since it connotes the more physically visualizable alteration of a molecular configuration.

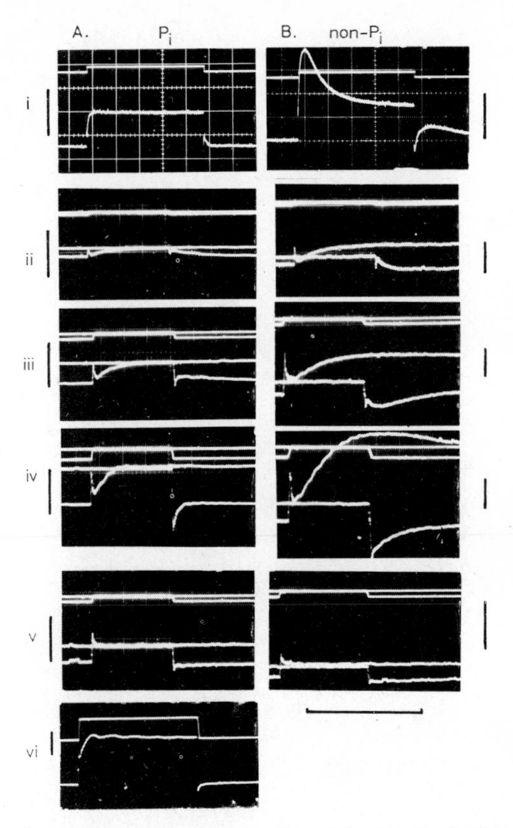

FIG. 43. Experimental responses of *Lethocerus* dorsal longitudinal flight muscle measured by White and Thorson (from White, 1973). Experimental conditions given in White and Thorson (1972). A, In the presence of 20 mM phosphate. B, In the absence of phosphate. Record (i) shows the response of tension (lower trace) to a 0.5% step length-change (upper trace) on a slow time scale. Records (ii)–(iv) show responses on a faster time scale to step length-changes of 0.2, 0.5 and 1%. In these figures the records to the step increase and step decrease have been superimposed by triggering the oscilloscope twice during the experiment. In records i–iv the muscle is activated by Ca^{++}. Record (v) shows the response of the relaxed muscle. Record (vi) shows the response to a step length change of 2% applied to the active muscle, illustrating the isometric oscillations. Vertical lines: 10 μdyn per A filament. The horizontal line corresponds to 5 sec (A (i)); 1 sec (B (i)); 500 msec (B (vi)); 100 msec (the remainder). Zero tension is approximately at the bottom of each record.

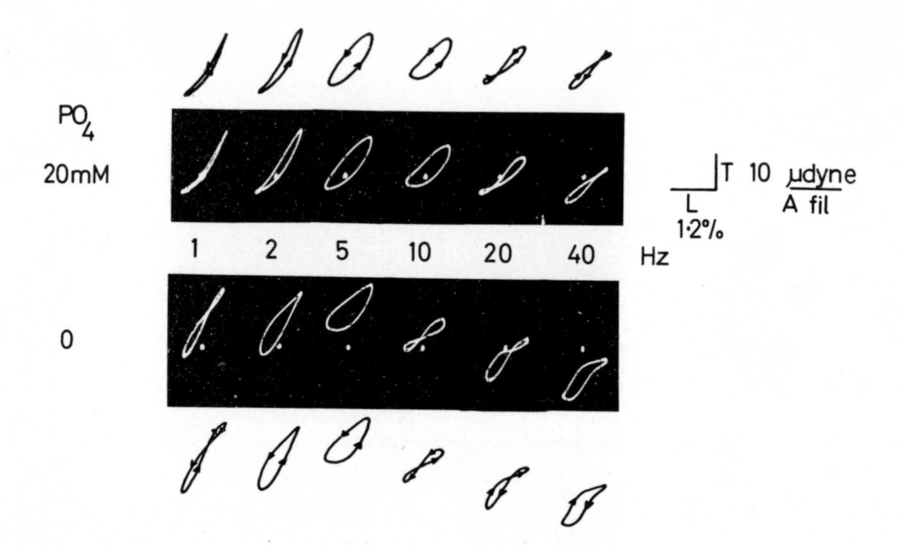

FIG. 44. Responses of the *Lethocerus* dorsal longitudinal flight muscle to sinusoidal length changes at the frequencies indicated, in the presence and absence of phosphate ions. The tracings adjacent to the records indicate the direction of rotation of the response. The tension in the absence of oscillation of length is illustrated by the dot. (From White and Thorson, 1972.)

tension change no longer rises monotonically to a new equilibrium level. Rather, the tension overshoots markedly before returning over seconds to a final equilibrium level—approximately that which would have been reached in the presence of phosphate for that particular step height. This we have called a "phosphate starvation transient" (PST).

Strikingly, without P_i, the response to the step decrease in length is not even qualitatively the inverse of that to the step increase, as is the case with P_i present. Provided the initial length is sufficient (at least 1–2% above rest length) there is a large PST-like *increase* in tension in response to the *decrease* in length (Figs. 43 B(i), iii, and iv). The PST remains even though the duration of the step change in length is made very small, and under these circumstances is in fact larger following a pulse decrease of length.

(iv) If the amplitude of the applied step length-change is greater than about 2% and less than about 6%, then the response has, superimposed upon that described in paragraph (iii), a series of damped oscillations (Schädler *et al.*, 1969, 1971; White and Thorson, 1972). In the absence of phosphate the period of the oscillation is much less than the duration of the PST.

(v) The response to applied sinusoidal length changes is more difficult to describe than that to step changes of length. At sufficiently low amplitudes (less than about 0.1% length change peak-to-peak) the resulting tension can be nearly sinusoidal (Jewell and Rüegg, 1966), and if linearity is assumed the response is specified fully in terms of amplitude and phase of tension with respect to length (i.e. dynamic stiffness) vs. frequency. Over a range of frequencies (between about 1 and 50 Hz in *Lethocerus cordofanus*) the tension is delayed behind the applied length changes, which means that the muscle is producing an amount of work each cycle equal to the area of the length–tension diagram; Abbott (1968, 1973) shows that the response obtained from the low-amplitude experiments is close to that expected from a simple exponential delay, corresponding to the exponentially delayed tension changes observed following a step change of length at low amplitude.

At higher amplitudes of applied sinusoidal length change the resulting tension is no longer a simple sinusoid (Pringle and Tregear, 1969; Thorson and White, 1972). Responses over a limited range of frequencies are shown in Fig. 44. Analysis of this response is difficult, and for kinetic studies step responses have been more satisfactory. However, for certain energetic studies the oscillatory measurement is natural since measurement of mechanical work obtained from the muscle is straightforward, and the muscle can be made to perform oscillatory work for long periods; stress relaxation is then ignored or obviated. The frequency of the maximum work per cycle at any amplitude of oscillation decreases as the amplitude increases. This frequency is approximately halved as amplitude is changed from 0.2% to 2.0% amplitude (peak-to-peak) of applied strain in both phosphate and non-phosphate solutions.

Lorand and Moos (1956) and Goodall (1956) demonstrated that glycerol-extracted rabbit psoas muscle could produce oscillatory work when mounted on a resonant lever, and this has been confirmed using forced sinusoidal oscillation of length by Rüegg *et al.* (1970), Heinl (1972—using frog sartorius), and ourselves (if P_i was present; unpublished observations). Aidley and White (1969) demonstrated similar properties in a neurogenically controlled cicada timbal muscle; they were unable to demonstrate any increase in ATPase activity with maintained stretch in the cicada muscle (nor in experiments in which the tension was kept high by slow extensions of the muscle as it stress-relaxed (unpublished observations by Aidley and White)). Both the rabbit psoas and cicada muscles produced delayed tension changes when subjected to step length-changes. (See also A. F. Huxley and Simmons, "phase 3", 1973.)

(vi) Rüegg *et al.* (1971) showed that the addition of orthophosphate ions to the solutions causes a reduction in the steady-state tension of the fibres, little change in the ATPase activity, and an increase in the speed of the response, measured either as the frequency of the isometric oscillations at high amplitude, or as the frequency at which the maximal amount of work per cycle is obtained from the fibres when the length is oscillated sinusoidally at a peak-to-peak amplitude of 2% (see also White and Thorson, 1972). Abbott (1973) found that the small-signal rate constant at low amplitude is also usually increased by the addition of P_i, although White and Thorson (1972) found no change. The P_i concentration required for half the maximal effect in all these experiments was between 1 and 5 mM P_i.

(vii) In the presence of applied sinusoidal length-changes the ATPase activity of the active fibres is further increased above that of the static fibres at the mean length of the oscillations, for frequencies of oscillation resulting in oscillatory work output from the muscle (Rüegg and Tregear, 1966; Rüegg and Stumpf, 1969b; Steiger and Rüegg, 1969). During the oscillation the mean tension of the fibres rises above that obtained in the absence of oscillations (see Fig. 44), especially in the absence of phosphate ions. Pybus and Tregear (1973) demonstrated that the ATPase activity of the fibres oscillated at the frequency giving maximum power output was approximately double that of the static fibres at the same mean tension, and up to three times that of the static fibres at the same mean length. The effect of oscillation upon the ATPase activity of the fibres is frequency-dependent, as shown in Fig. 45. The increase in ATPase activity is approximately proportional to the power output in experiments in which the mean length of the fibres is held constant and the frequency of the oscillation varied (Steiger and Rüegg, 1969). They called this observation the biochemical equivalent of the Fenn effect. Pybus (1972) showed that the linear relationship between ATPase activity and power output broke down when the power output at any oscillation frequency was reduced by increasing the tension in the muscle sufficiently. Under these conditions the ATPase activity is about linearly related to the mean tension in the fibres.

(viii) As Pringle suggested, the ability of the muscle to produce work during sinusoidal oscillation (and hence, presumably, during partially synchronized molecular changes) makes the insect flight muscle an excellent candidate for X-ray diffraction measurements synchronized to particular phases of the cycle.

Armitage *et al.* (1973) summarize the work that has been done along these lines. They have used a proportional counter to monitor intensity of particular portions of the diffraction pattern at a given phase of the cycle, and thus derived the changes in intensity during the cycle of oscillation. Data from many cycles must be averaged to produce statistically significant results. Cyclic intensity fluctuations for active muscle can then be compared with those observed in relaxed and rigor muscles, and with those for active muscle in the absence of oscillation. So far results have been obtained from the (1,0) and (2,0) equatorial reflections, from the 14.5 nm meridional and from the 38.5 and 19.3 nm layer lines.

Miller and Tregear (1972) have presented a detailed interpretation of the X-ray patterns. Briefly, the changes in intensity of the two equatorial reflections is related to the transfer of mass between the A and I filaments, and the intensity of the 14.5 nm spot gives information about the axial orientation of the cross-bridge (presumably the angle between the axis

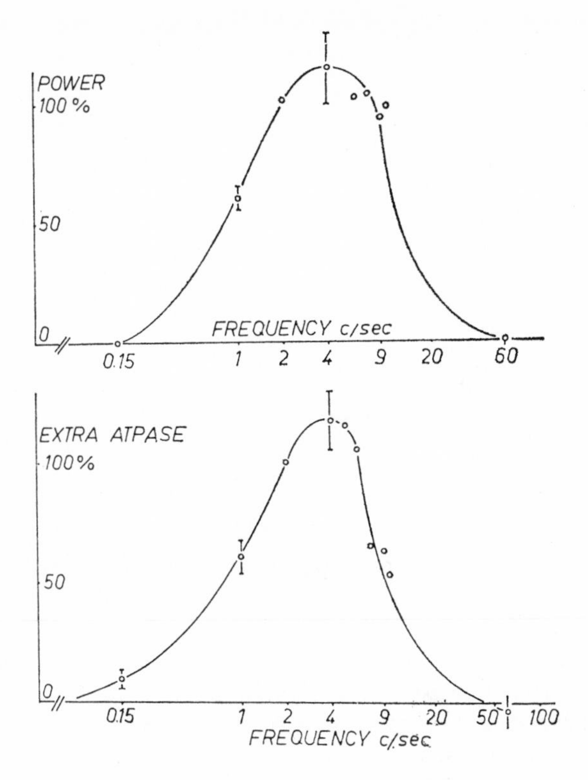

FIG. 45. Dependence of power and "Extra ATPase" (i.e. ATPase activity in excess of that obtained in the absence of oscillation) upon frequency for applied sinusoidal length oscillations of 2.5% peak-to-peak amplitude (Steiger and Rüegg, 1969).

of the S_1 and the filament axis). The interpretation of the two layer lines is less clear because the helical repeats of (1) cross-bridge origination sites on the A filament, and of (2) actin monomers on the I filament are both 77 nm. Thus both can contribute to the 38.5 and 19.3 nm layer lines. In particular the cross-bridges can contribute to these layer lines in two ways: (1) in relaxed muscle by marking the myosin helix, and (2) in rigor muscle by marking the actin helix.

The intensity changes during oscillation have been measured at the frequency of optimal work per cycle (about 4 Hz, in the absence of phosphate) for all of the above reflections, and at 1.2 and 18 Hz for the 14.5 nm reflection. The changes in intensity observed were always nearly in phase (or in antiphase) with the tension exerted by the muscle, rather than with the length changes applied. The observations are summarized in Table 4.

In fact in the above phase measurements a small lag in intensity with respect to tension was generally observed. If the intensity changes are in phase or antiphase with bridge attachments, a lag is expected on the basis of our formulation (Thorson and White, 1969); that is, measured oscillatory tension is the resultant of both nearly elastic (in phase with length) and cross-bridge (lagging) forces, and so must have an intermediate phase.

The cyclic changes in intensity of the equatorial reflections were so small as to be not statistically significant. However, they were dependent upon the Ca^{++} concentration and suggested as one interpretation that an increasing proportion of the cross-bridges move to

TABLE 4. RELATIVE INTENSITIES OF X-RAY DIFFRACTION REFLECTIONS UNDER DIFFERENT MECHANICAL CONDITIONS

State of muscle	Reflection				
	14.5 (%)	38.5 (%)	19.3 c.p.s.	(1,0) (%)	(2,0) (%)
1. Relaxed	100	100	0	100	100
2. Ca-activated	100	(100)	(0)	103	108
3. Oscillatory					
a. Min. tension	98	(100)			
b. Max. tension	76	81		pp 2.6[a]	pp 4.2[a]
4. Rigor	12	353	1.9	71	162

[a] These are the peak-to-peak changes in intensity observed on oscillation. Their absolute values on the intensity scale are not available. Values in parentheses are approximate. (From Miller and Tregear (1972) and Armitage *et al.* (1973).)

the vicinity of the I filament as the Ca^{++} concentration is increased. The lack of significant intensity changes in the other reflections upon Ca^{++} activation (in the absence of oscillation) perhaps suggests that the above bridges do not attach to the actin monomers. The tension exerted by the fibres during the "static" measurements was very low (Tregear, personal communication), so that no activation of the static fibres occurred due to applied strain. It would be of great interest to know what intensity changes occurred with changes in static tension (the experiment could be done at an oscillation frequency of, say, 0.01 Hz if necessary to obviate the effects of stress relaxation).

The effect of oscillation (which produces tension changes in the fibres from close to zero tension to large values under the conditions used) is to cause little change in the "fraction of cross-bridges in the vicinity of the I filaments" (as inferred from the equatorial reflections), but large changes (of the order of 10–20%) in the "fraction of cross-bridges attached" (as inferred from the intensity changes of the 38.5 and 19.3 nm reflections). Miller and Tregear's interpretation implies that virtually none of the bridges is attached when the tension is near zero, but that most of those in the vicinity of the actin are attached when the tension is maximal. The large intensity changes observed in the 14.5 nm reflection could be accounted for if the attached cross-bridges are all at the rigor angle (rather than being distributed in angle from the rigor angle to the perpendicular); the intensity changes can be explained by about 30% of the cross-bridges having an axial displacement of 2 nm (Tregear and Miller, 1969); if less than 30% of the cross-bridges are attached (as inferred from the other reflections), and if only attached cross-bridges are displaced axially by the oscillation, then the axial displacement must be large. Armitage *et al.* (1973) suggest that this situation is compatible with the notion that the S_2 link between the backbone of the A filament and the S_1 head is compliant, allowing the attached cross-bridge to change its angle in the absence of relative filament sliding.

3. *The Relationship of Insect Fibrillar Flight Muscle to Vertebrate Striated Muscle*

Despite the differences between the two types of muscle—the maintained strain activation, high resting stiffness, arrangement of filaments in the sarcomere and of molecules within the A filament—the hope has been that comparative studies would provide clues for an understanding of vertebrate muscle.

In particular, as Professor J. W. S. Pringle first realized, the cyclic performance of work by the insect muscle raises several experimental possibilities, especially the study of the structural events underlying contraction by means of X-ray diffraction and of the relationship between ATP hydrolysis and energy output. So far the results most applicable to vertebrate muscle have been the clues from structural correlates of activity. The structural analysis of relaxed and rigor muscle in terms of cross-bridge conformation changes (Reedy *et al.*, 1965; Reedy, 1968) and the continuing study of the changes taking place during the oscillatory cycle (Miller and Tregear, 1972) are giving a more detailed picture of the working of the cross-bridge cycle than is being obtained from the vertebrate striated muscle.

However, realization of the original hope has proved elusive, in particular because the mechanism of strain activation has not yet been discovered. If this were determined, and were found to act by a change in one or more of the rate constants of the biochemical cycle, then insect fibrillar flight muscle would be particularly useful material for experiments relating the biochemical and mechanical cycles.

The static strain-activation properties of the insect muscle have resulted in qualitative differences between models of cross-bridge activity for insect fibrillar muscle and those for vertebrate muscle. These are made explicit in the next section.

4. *Cross-bridge Analyses of Insect Fibrillar Flight Muscle*

During 1966, in Pringle's laboratory and at his suggestion, we examined the relationship of A. F. Huxley's cross-bridge cycle to the mechanical behaviour of *Lethocerus* muscle. The critical mechanical property which enables the work production in flight is that imposed length changes induce delayed tension changes. Similarly, in any hypothetical cycle of attachment and detachment, the fraction of bridges attached must undergo delayed readjustment following alteration of the rate constants. We naturally proposed, therefore, that the critical delay in tension was the delay in readjustment of this attached fraction (Thorson and White, 1969).

In order to test this idea quantitatively we had to make assumptions as to the means by which stretch perturbed the rate constants. Since fibrillar muscle is very stiff, and A-filament-to-Z-line links are suspected (see § III), it was natural to assume that the (necessarily maintained) perturbation was mediated by strain in the A filaments.

The exponentially delayed tension arises naturally from the simplest equation for the two-state cycle, in which n denotes the fraction of attached cross-bridges,

$$\frac{dn}{dt} = f(1 - n) - gn,$$

for if one treats the rate constant f as a forcing function, with g constant and $f \ll g$ so that $n \ll 1$, the solution in the Laplace variable s is

$$n(s) = \frac{f(s)}{s + g}.$$

This transfer function is simply that of an exponential-delay (first-order) process, with rate constant g. This notation is useful since, as Machin and Pringle (1960) showed, one can actually do small-signal sinusoidal analysis of this muscle and approximate the above conditions.

More important, however, is the fact that with the extremely small oscillations of length practicable experimentally (*ca.* 0.05% peak-to-peak; Abbott, 1968) attached cross-bridges cannot undergo appreciable distortion. Hence the unknown dependencies of bridge force and g upon distortion, severe sources of arbitrariness in all large-signal analyses of the cyclic hypothesis, did not affect our tests.

Two critical tests of the above idea were feasible at that time. First, we were concerned that when cycling bridges were embedded properly in the multi-region viscoelastic sarcomere, the above neat explanation could fail. This is a particular worry with strain activation because there is positive feedback between the attached fraction of bridges and strain, so that the bridges effectively interact. As indicated diagrammatically in Fig. 39, there is also a distribution in the stress (and thus in the strain) along the filaments. If there is any effect of strain upon the activity of the cross-bridges this will therefore be different at different points along the filament. White (1967) had characterized the mechanics of the passive sarcomere sufficiently that we were able to test this worry, calculating the length–tension dynamics of a representative non-homogeneous viscoelastic sarcomere with strain-activated cycling bridges in the overlap zone. Not only did the predictions agree well with Abbott's (1968) small-signal data (Fig. 46), but a further independent test—comparison of predicted cycling rate of bridges with ATP hydrolysis measurements—produced striking agreement (Thorson and White, 1969).

Other distinct suggestions for the origin of the work production current at that time (Jewell and Rüegg, 1966; Pringle, 1967; Abbott, 1968) have apparently not survived as fundamental alternatives (but see paragraphs c and d below) although the variables invoked (Ca^{++} binding, cross-bridge distortion, and ADP diffusion) are by no means to be ignored.

Work during the intervening 6 years has altered our view of the cyclic bridge hypothesis for fibrillar muscle considerably. We shall outline recent developments as they have influenced that view.

a. *The small-signal case*

Abbott (1973) has studied the small-signal sinusoidal response of bundles of ten freshly glycerinated fibres, paying particular attention to the time elapsed since glycerination. There appear to be distinct advantages in using freshly glycerinated fibres (cf. our discussion above of live fibres), which have less Ca^{++} sensitivity than after long storage but seem more reproducible. He finds that the small-signal rate constant for delayed tension is independent of the degree of steady stretch, but that the delayed tension itself increases with stretch (in the range 0–2.5%). It appears likely (Abbott, personal communication) that the latter effect is accounted for by the nonlinear dependence of relaxed stiffness upon stretch in this preparation, since the latter varies with stretch in a manner like that of the delayed component. The meaning of the effect is therefore not yet clear.

Abbott also finds that the Ca^{++} concentration in the range pCa 6.5–5.6 *increases* the rate constant for the delayed tension. This is hard to reconcile theoretically with his result (Abbott, 1968; and also in the freshly glycerinated bundles) that Ca^{++} reduces the general stiffness of the fibres at all frequencies (i.e. the Nyquist plot shifts to the left), for decreased stiffness of the filaments tends in our analyses (Thorson and White, 1969) to *decrease* the measured rate constant. It seems most significant, therefore, that in A. F. Huxley and Simmons' (1971a, b) formulation, decreased stiffness of such structures as the S_2 link in

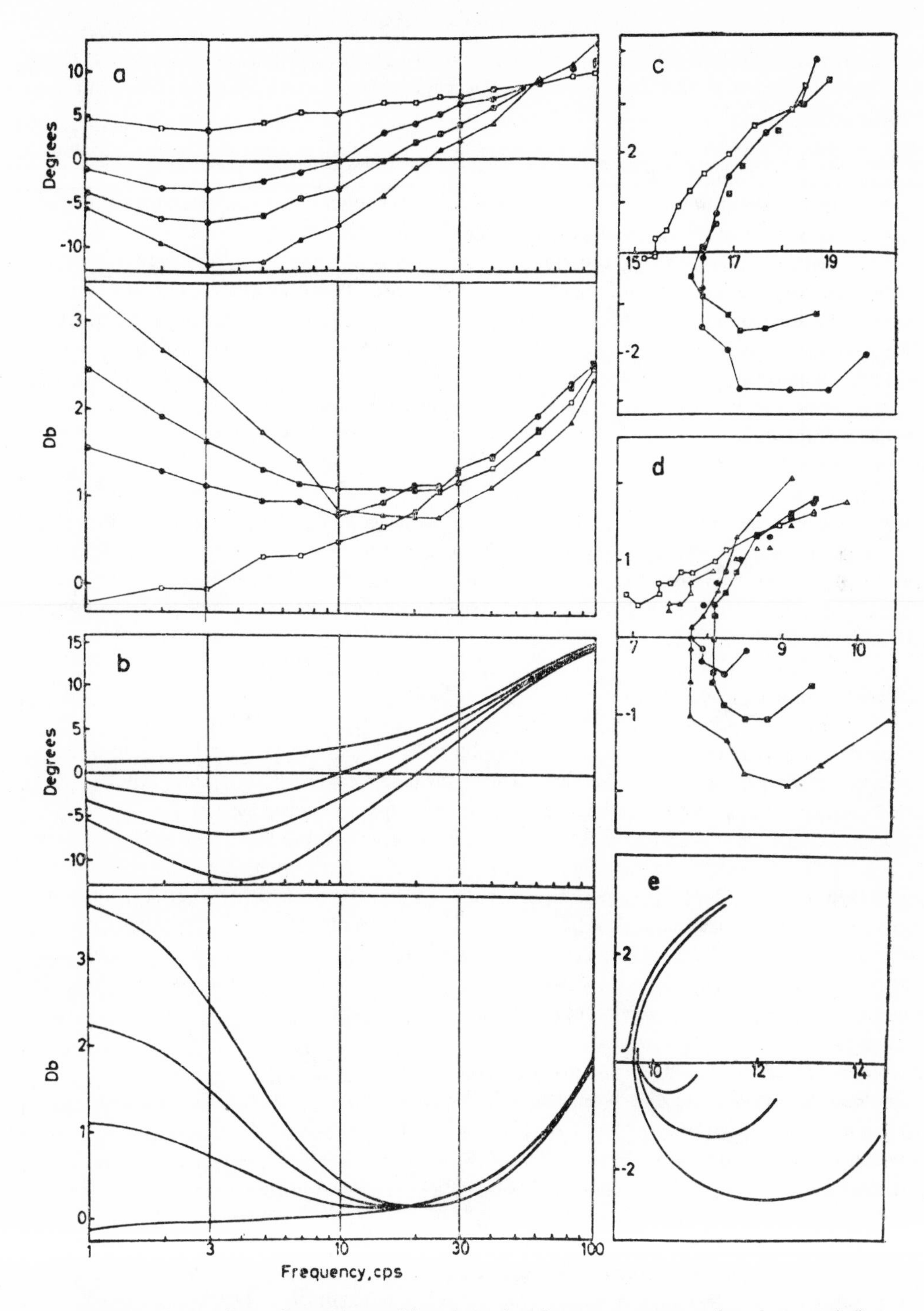

FIG. 46. Comparison of experimental length-tension frequency–response data (a, c, and d) (Abbott, 1968) and the predicted response of the non-homogeneous viscoelastic half-sarcomere (b and e) from Thorson and White (1969). a and b: Bode plots of amplitude and phase against frequency. c, d, and e: Nyquist plots of the in-phase and quadrature components of the dynamic stiffness. Filament elasticities assumed are those of Table 3.

fact *increases* the rate constants for transitions among attached states. This might account for Abbott's result if the delayed tension were associated with Huxley and Simmons' transitions.

b. *Nonlinearity of the two-state cycle*

Since n saturates at unity as $f \to \infty$, and since the rate constant for approach to equilibrium is $f + g$, the two-state cycle is inherently nonlinear for large changes of f.

It sometimes helps to know the effects of various nonlinearities alone before they are treated in concert. The individual effects will not add, but they often combine without mutual annihilation. To this end, we (Thorson and White, 1969) showed that the clamshell-like length–tension loops at large length perturbation arise in n vs. f loops from special perturbations of the two-state cycle, and pointed out that nonlinear strain activation of f could improve the description of the tension–length nonlinearity. Abbott (1973), also treating only the elementary cycle, confirms that the latter assumption produces n vs. f clamshells as well.

In the same spirit one can examine the effect of allowing bridge distortion to change interfilament shear force in the absence of its suspected effects upon g. One then finds (White and Thorson, 1972) that two of the nonlinearities of fibrillar muscle in P_i (both (a) early elastic tension changes and (b) the rate constant for the delayed component of tension, depend upon the tension at which a large length step is imposed) are reproduced qualitatively.

c. *Julian's interpretation*

Julian (1969) offered a description of the delayed-tension properties of insect flight muscle in terms of a two-state cross-bridge cycle in which the attachment probability f is dependent upon strain as in our proposal; however, following a suggestion of Jewell and Rüegg (1966) *re* delayed Ca^{++} binding, he included an exponential delay limiting the changes of f, implying that the delayed tension changes then predicted are due to the delayed change in f. In order to sort out these effects, we repeated Julian's calculation without his delay between strain and f (White and Thorson, 1972) and demonstrated that the basic delay in Julian's calculation remains, in accordance with attachment and detachment rates as in our earlier formulation (Thorson and White, 1969). Hence, though a limiting delay between strain and f has not been ruled out, there is no demonstration that it is either necessary or sufficient for the work-producing properties of fibrillar muscle.

d. *Model of Schädler, Steiger, and Rüegg* (1971)

In their paper treating the isometric oscillation of *Lethocerus* muscle, Schädler *et al.* (1971) discuss a qualitative model for the operation of insect muscle. This model relies heavily upon their observation that the "instantaneous-elasticity" (the magnitude of the initial rise in tension following a step length change of constant amplitude) does not change during the rising phase of the delayed tension changes. These observations were made by applying a small test step length-change at various times after the application of a *very* large (6%) initial length-change. We have done a similar experiment (unpublished experiments), but with the test steps applied after a much smaller initial length-change ($\frac{1}{2}$ or 1%), and find that the instantaneous stiffness increases during the rising phase of the delayed tension change.

On the other hand, the notion in the model of Schädler *et al.* that the delayed tension following stretch occurs during changes of state of attached bridges is a stimulating alternative to the one we have tested and, formulated nonlinearly, need not stand or fall on the above measurements of instantaneous stiffness. Moreover, if these attached states were to be related as in Huxley and Simmons' (1971a, b) scheme, one may be able to account for Abbott's (1973) Ca^{++} effects, as we have described in (a) above.

Schädler *et al.* (1971) also claim that "transient cross-bridge synchronization obviously gives rise to the tension overshoot and to the isometric oscillation", but offer no explanation of this proposal in terms of the recognized variables of current cross-bridge theory. As we have emphasized above and in Appendix II, we have not yet found even a qualitative basis for the isometric oscillations which also accommodates the other known properties of flight muscle.

e. *Three-state cycles*

Despite the above mathematical experiments, no one has suggested a way in which the two-state cycle might be modified plausibly to account for such remarkable nonlinearities (listed above) as the curious phosphate starvation transients (PSTs) and the "resonance" of ATPase rate as large-signal oscillation frequency is changed.

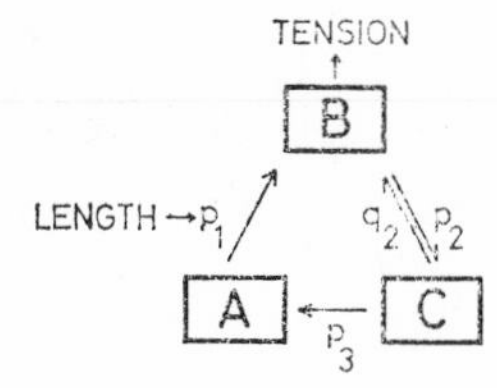

FIG. 47. The three-state cross-bridge cycle of White (1973).

White (1973) has shown that, rather surprisingly, the PSTs are well reproduced by the three-state cycle of Fig. 47. Only state B is tension producing. In order to produce the PST-like transient responses of Fig. 43 A it was assumed that p_1 was proportional to $(\text{length})^2$, and that p_3 was distortion dependent. The rates of steps p_2 and q_2 were both assumed to be increased by inorganic phosphate, q_2 conceivably corresponding to binding of P_i.

We have since found that the proper general form of the relationship between ATPase activity and oscillation is also given by this three-state cycle. In order to obtain the maximum amount of mechanical energy per cross-bridge cycle it is desirable that the detachment rate (p_3 in the three-state cycle) be very much larger for that value of distortion at which the force exerted by a bridge is close to zero than for distortions where the force is large and tending to produce shortening.

The responses illustrated in Figs. 48 and 49 were obtained with p_3 and bridge shear force dependent upon distortion as shown in Fig. 50. The responses to step length-changes show the PST on both the step-up and the step-down. Moreover, the sinusoidal waveform shows many of the features observed experimentally, including the tendency to produce figures-of-eight (Fig. 44) at rather low frequencies, and also for the average tension during the cycle to rise during oscillations at frequencies giving oscillatory work. Figure 51 shows the average cycling rate vs. frequency for the responses illustrated in Fig. 49. There is a large increase in "ATPase" (i.e. cycling rate) over the working range of frequencies.

The response of the three-state scheme at high frequencies, however, does not become

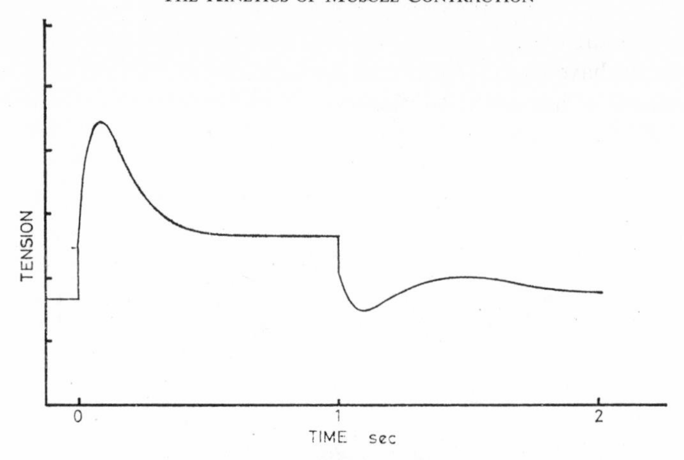

FIG. 48. Response of the three-state cycle to a step change in strain from 3.2 to 7.1 length units. The dependencies of p_3 and tension upon distortion used are as shown in Fig. 50. Tension units arbitrary. $p_2 = 12\ \text{sec}^{-1}$; $q_2 = 0.2\ \text{sec}^{-1}$; $p_1 = (\text{strain})^2/5$.

energy absorbing (clockwise cycling of the tension–length loop). Nor are isometric oscillation seen. Some methods for seeking this latter property are included in Appendix II. Our view at the moment, however, is that adequate cross-bridge models of fibrillar muscle have become sufficiently arbitrary that one cannot foresee clear exclusion of alternatives without further experimental constraints. Nonetheless, both the intuition gained from these early efforts to understand the events underlying contraction and the prospect of deciding among specific alternative ideas add special relevance to questions now being posed: Are isometric oscillations truly (by spot follower standards) isometric? Are PSTs specific to lack of the PO_4 ion? Can tryptophan fluorescence reduce the ambiguity of inference from X-ray diffraction? How much complexity (especially *re* stress relaxation) is introduced by glycerination? What in fact are the chemical correlates of strain activation?

VIII. LINKS BETWEEN THE MECHANICAL AND BIOCHEMICAL KINETICS

In § IV we presented the current evidence concerning the sequence of reactions by which ATP is hydrolysed by actin–myosin systems in solution, and summarized, in Fig. 20 B, a possible scheme for the main cyclic sequence. In § VI and VII, on the other hand, we described recent ideas about the mechanical states through which a cross-bridge is considered to go as it contributes cyclically to interfilament shear force. It is generally hoped that a mapping between the mechanical and biochemical cycles can be discovered. In the meantime, candidate mappings can suggest decisive experiments.

The main difficulty, of course, is that in intact muscle the actin and myosin molecules are subjected to mechanical forces and constraints not present in solution. Indeed, contraction itself is thought to depend upon these forces, and it is these which define the putative states (attachment, detachment, degree of distortion) of the mechanical cross-bridge cycle. We need not expect to find precisely the same molecular species, much less similar rate constants, in the two cycles. For this reason, quick quantitative comparisons (e.g., of maximum ATPase in actomyosin gels with maximum speed of shortening of fibres, or of supposed rate-limiting steps in the two cycles) can be premature, if not actually misleading.

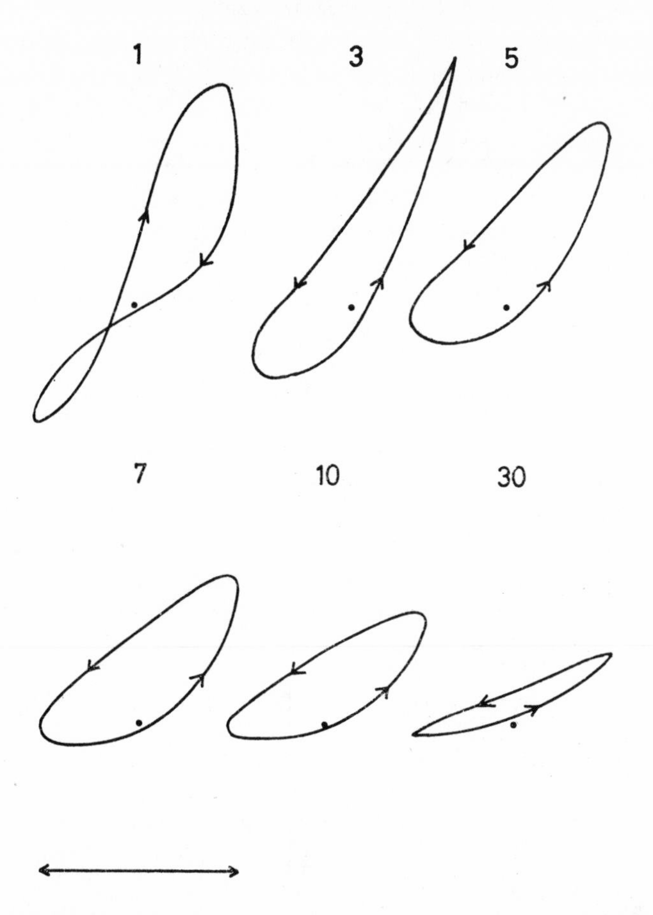

FIG. 49. Tension-length loops of the three-state cycle in response to sinusoidal length changes at the frequencies indicated, under the same conditions as those used to obtain Fig. 48. The applied sinusoidal strain had an amplitude of 4 length units, superimposed upon a steady strain of 5 length units; as in Fig. 48, p_1 = (strain)^2/5. The steady-state tension (in the absence of oscillation) is indicated by the dot.

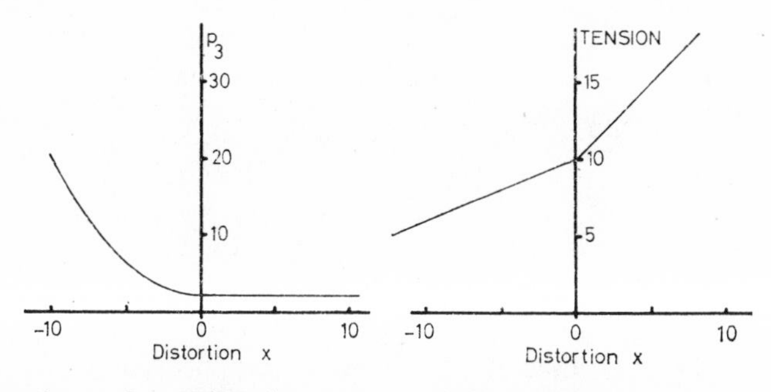

FIG. 50. Dependence of p_3 and tension upon cross-bridge distortion used to obtain Figs. 48 and 49. The reason for introducing the nonlinear tension vs. distortion curve, discussed by White (1973), is to accommodate the reduced instantaneous stiffness of the fibres to a step decrease in length.

Qualitative comparisons, however, are compelling. As we have stressed, an apparent peculiarity of actin–myosin ATPase is that in binding ATP so as to split it, actomyosin is itself dissociated. This situation has been compared in the mechanical cycle with the binding of ATP by an attached bridge, detachment ("dissociation") and the splitting of ATP in the detached state. This idea is qualitatively compatible with the known "plasticizing" effect of ATP and with the ATP dependence of the flight-muscle mechanical rate constants (White and Thorson, 1972). One might then compare the "detached" bridges of relaxed muscle with the M*ADP·P state of Fig. 20 B; in fact, Marston and Tregear (1972) have demonstrated that in relaxed glycerol-extracted rabbit psoas muscle the predominant state of the myosin is as a myosin–ADP complex (whether the inorganic phosphate P_i was also bound was not tested).

Several distinct approaches to the biochemical identity of the contractile events now offer promise:

(1) It is possible biochemically to "stop" the cycle in intact fibres at a known point, and then to determine the mechanical state and structure of the cross-bridge. One such state is the "rigor" state, which is obtained in the absence of any nucleotide, and in which the cross-bridges appear to be angled at about 45° (Reedy *et al.*, 1965; see also Fig. 13). The biochemical state is possibly a correlate of the state "AM" of Fig. 20 B. Another such state is obtained in relaxed muscle, as discussed above.

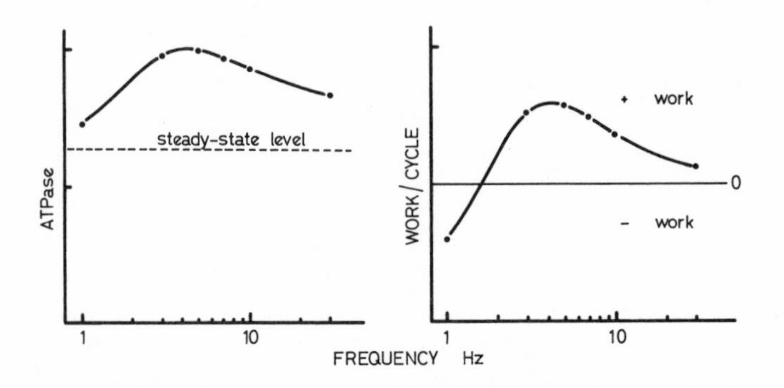

FIG. 51. The ATPase activity and work/cycle obtained from the three-state cycle under the conditions of Fig. 49. The ATPase curve was obtained by assuming that the ATPase activity was instantaneously proportional to p_1 × (fraction of cross-bridges in state A). ATPase and work/cycle units are arbitrary.

Using ATP analogues one may be able to stop the cycle at other points and then to determine the structural and mechanical properties. For example, White (1967, 1970) demonstrated that the mechanical properties of muscle "relaxed" with pyrophosphate were different from those of ATP-relaxed muscle; it is possible that the state obtained corresponds to the M·ATP (or perhaps the AM·ADP) state of Fig. 20 B.

(2) By the use of absorption, fluorescence, or other forms of spectroscopy it may become possible to monitor identifiable changes among biochemical states in intact fibres. The strain-activation properties of insect fibrillar flight muscle makes it an obvious candidate for such experiments, as emphasized in § VII.

(3) The biochemical cycle includes steps in which the ATP is bound, and ADP and P_i are released; by changing the concentrations of these substances the rates of the steps

involving their binding (i.e. the forward step of ATP binding, and the reverse reactions for the ADP and P_i release) should be changed. The effects on the mechanical properties of the muscle should thus give clues as to the role of these steps in the biochemical cycle. Some preliminary experiments along these lines are reported in § VII; we have shown that the role of free phosphate may be more complex than that of slowing only the release of bound phosphate.

(4) "Mathematical experiments" examining the consequences of alternatives within cross-bridge theory can now be done with close coupling to a steady supply of new results on structure, solution chemistry and the mechanical dynamics. For example, the separate indications from structural (H. E. Huxley and Brown, 1967; Armitage *et al.*, 1973) and biochemical (Eisenberg and Kielley, 1973) experiments that only a small fraction of the active bridges are attached at any instant have specific consequences in the theoretical formulations. Moreover, there is much room for improving our computational techniques to handle more efficiently the full nonlinear partial-differential equations of the sarcomere.

At the moment, then, there is relatively little evidence to match the states of the mechanical cycle with those of the biochemical cycle. But work along the above lines will surely be the basis of exciting advances during the next decade.

APPENDIX I.

BIOCHEMICAL KINETICS

The following is a brief analysis of the kinetics discussed in § IV on Biochemical Kinetics.

Consider the simple reaction

$$\mathrm{A} \underset{k_{-2}}{\overset{k_2}{\rightleftharpoons}} \mathrm{B}. \qquad (\mathrm{A}.1)$$

We have called the rate constants k_2 and k_{-2} in order to cause less muddle later.

Suppose one starts with a concentration A_0 of A and no B, and wishes to follow the way in which the system reaches equilibrium. At some time t after the reaction has been started the concentrations are $A(t)$ and $B(t)$; the rate of the forward reaction is $k_2 A(t)$ and that of the backward reaction $k_{-2} B(t)$. The net rate at which B is being formed is the difference between these:

$$\frac{dB(t)}{dt} = k_2 \cdot A(t) - k_{-2} \cdot B(t).$$

The total number of molecules in the solution is constant and equal to the starting concentration A_0. Hence $A(t) = A_0 - B(t)$ and

$$\frac{dB(t)}{dt} = k_2 \cdot (A_0 - B(t)) - k_{-2} \cdot B(t)$$

$$= k_2 \cdot A_0 - B(t)(k_2 + k_{-2})$$

with the solution, for the case $B(0) = 0$,

$$B(t) = \frac{k_2 A_0}{k_2 + k_{-2}} [1 - \exp(-(k_2 + k_{-2})t)]$$

(for if this is differentiated one obtains the above differential equation). Thus the system approaches equilibrium with a rate constant $k_2 + k_{-2}$, the sum of the forward and backward rate constants. (The rate constant of the exponential curve is independent of the starting concentration A_0, though the absolute conversion rate at any time is not.)

Now consider the more complicated reaction

$$M + S \underset{k_{-1}}{\overset{k_1}{\rightleftharpoons}} MS \underset{k_{-2}}{\overset{k_2}{\rightleftharpoons}} MP.$$

In the stopped- and quenched-flow experiments, M and S are in the starting syringes, and the formation of MP is measured. Certain limiting conditions are of interest:

1. *Excess S.* Consider the first reaction in isolation. If at the start $S_0 \gg M_0$, S is nearly constant and one can write that the equilibrium $M/M_0 = k_{-1}/(k_1S + k_{-1})$ is approached with the rate constant $(k_1S + k_{-1})$. If, further, S is so large that $k_1S \gg k_{-1}$, nearly all the M is removed from its free state and this situation is reached very rapidly. If now both reactions are considered, the fact that $MS(t) < M_0$, independent of the values of k_2 and k_{-2}, means that the above considerations still apply. Finally, therefore, if $k_1S + k_{-1} \gg k_2 + k_{-2}$, the rate constant for the isolated second reaction, MP will accumulate nearly exponentially with the measured rate constant $k_2 + k_{-2}$, until $MP/MS = k_2/k_{-2}$. That is, injection of the M_0 and S_0 is indistinguishable from the injection of MS in an amount equal to M_0.

2. *Low concentration of S.* If S_0 is sufficiently small that $k_1S_0 \ll k_2 + k_{-2}$, the behaviour then depends upon k_{-1}: (a) $k_{-1} \ll k_2 + k_{-2}$. Here the ratio MP/MS is never far from its equilibrium value k_2/k_{-2} ($\equiv K_2$), which is attained, with the rate constant $k_2 + k_{-2}$, more rapidly than step 1 can perturb it. If $S_0 \gg M_0$ despite the above restriction on S_0 (otherwise the kinetics are nonlinear and quite complicated), S is nearly constant and equal to S_0. Hence, with $M(t) = M_0 - MS(t) - MP(t) = M_0 - MP(t)\,(1 + K_2)/K_2$, the differential equation for MP is

$$\frac{dMP(t)}{dt} = \frac{k_1 K_2 S_0 M_0}{1 + K_2} - \left(k_1 S_0 + \frac{k_{-1}}{1 + K_2}\right) MP(t)$$

so that MP accumulates exponentially with the rate constant $k_1S_0 + k_{-1}k_{-2}/(k_2 + k_{-2})$. (b) $k_{-1} \gg (k_2 + k_{-2})$. Here the equilibrium ($MS/M = k_1S/k_{-1}$) of the first stage of the reaction is reached very rapidly and maintained during the slower formation of MP. Again with $S_0 \gg M_0$,

$$MS = \frac{k_1S}{k_{-1}} M$$

so that

$$\frac{dMP(t)}{dt} = \frac{k_1 k_2 M_0 S_0}{k_{-1} + k_1 S_0} - \left(\frac{k_1 k_2 S_0}{k_{-1} + k_1 S_0} + k_{-2}\right) MP(t).$$

But since $k_1S_0 \ll k_2 + k_{-2} \ll k_{-1}$, the rate constant for accumulation of MP is simply $(k_1k_2/k_{-1})S_0 + k_{-2}$.

3. *Intermediate concentrations of S such that* $k_1S = (k_2 + k_{-2})$. (a) $k_{-1} \ll (k_2 + k_{-2})$. Here k_{-1} can be ignored, since this back reaction is slow compared both with the forward reaction and with the rate at which equilibrium is reached in step 2. The rates of the two stages of the reaction are equal so that MP builds up in a sigmoidal manner rather than

exponentially. (In the two previous cases one or other of the stages was so much faster than the other that the "lag" due to the second-order kinetics was not expected to be measurable.) If $k_1 \cdot S = k_2 + k_{-2} = a$, then (Bagshaw *et al.*, 1973) MP(t) is proportional to $(1 - e^{-at} - t\,e^{-at})$. (b) $k_{-1} \gg k_2 + k_{-2}$. This case has been treated in case 2(b) above, since the restriction $k_1 S \ll k_{-1}$ was applied only in the final equation. The corresponding rate constant is $(k_1 k_2 \cdot S)/(k_{-1} + k_1 \cdot S) + k_{-2}$.

APPENDIX II

SOME PROPERTIES OF THREE-STATE KINETICS

Several current analyses call attention to kinetic hypotheses in which more than two cross-bridge states (attached and detached) are involved. These include A. F. Huxley's and Simmons' (1971a, b) several states of attachment, their and Julian's (1973) demonstrations that these can describe certain mechanical events, White's and Thorson's (1972), and White's (1973) analyses of the relations of the cycle to the involvement of P_i, ATP, and ADP, as well as Descherevsky's (1968) formulation of the cycle.

We ought, therefore, to outline briefly certain general properties of this class of ideas. Consider the three-state scheme of Fig. 47, which includes one illustrative back reaction. Here we shall ignore entirely the plausible properties (a) that cross-bridge distortion (a function of both "time-since-attachment" and sliding of the filaments) ought to affect both tension and certain of the rate constants, and (b) that viscoelastic properties of the sarcomere (via which the bridge cycles may in fact interact (Thorson and White, 1969)) can be powerful. Even with these drastic simplifications, justified only by the aid to intuition which can come from treating a reduced model, the three-state case is qualitatively far more complex than the two-state one.

For example, consider the small-signal transfer function relating small perturbations δf of the "attachment" rate constant f to small fluctuations δA of the contents of the state A, in the two-state cycle ($D \underset{g}{\overset{f}{\rightleftharpoons}} A$). This is simply

$$\frac{\delta A(s)}{\delta f(s)} = \frac{g}{(f_0 + g)(s + f_0 + g)}$$

where $A + D = 1$, f_0 is the unperturbed or average value of f, and s is the Laplace variable. For the three-state case ($D \overset{f}{\rightarrow} A \underset{h}{\overset{e}{\rightleftharpoons}} B \overset{g}{\rightarrow} D$) with back reaction h, however, this becomes

$$\frac{\delta A(s)}{\delta f(s)} = \frac{eg}{(ef_0 + eg + f_0 g + f_0 h)} \frac{s + g + h}{s^2 + (e + f_0 + g + h)s + (ef_0 + eg + f_0 g + f_0 h)} \quad [22].$$

Not only do the specific chemical rate constants now interact in a complicated way with those of the cycle as a whole, but two further properties are explicit:

(1) Although the absolute net rate of transfer of bridges (or, more precisely, of probability density) from states A to B is slowed by the postulation of the back reaction h (as discussed by White and Thorson, 1972), the apparent rate constant for attainment of equilibrium (the equilibrium itself is much altered by the back reaction) has in this particular case been increased. Study of particular cases resolves such superficial anomalies, but the warning to the generalizer is clear.

(2) The second-order denominator of the transfer function accommodates overshoots and oscillations, given suitable values of the rate constants. This denominator is precisely the polynomial of the "characteristic equation" for the three-state system which could also be obtained by Liapounoff analysis of the corresponding set of differential equations,

$$\frac{dA}{dt} = f(1 - A - B) - eA + hB,$$

$$\frac{dB}{dt} = eA - (h + g)B.$$

For large changes of parameters such as f, the system is effectively nonlinear, and the simplest means of analysis is to integrate the simultaneous equations numerically on a digital computer. White (1973) illustrates this procedure in his analysis of the conditions under which such three-state systems can suffice for phosphate-starvation phenomena (White and Thorson, 1972). Moreover, he shows that there are rather strong interconstraints among the parameters e, f, g, and h if one is to describe both the peculiar dynamics and the measured steady states. He finds that the set 12, 6, 2, and 0.2 sec^{-1}, respectively, suffices approximately for non-P_i conditions.

With these constraints, a highly informative calculation can be made. Tension in insect fibrillar flight muscle "rings", or shows several cycles of (presumably) isometric damped oscillation, when the muscle is subjected to sudden large changes of length. Since the frequency of the ringing is temperature-dependent, it is conceivable that the cross-bridge cycle itself may be involved.

Furthermore, the above second-order characteristic equation of the three-state system, which (from the denominator of eqn. (22)) can be written

$$s^2 + 2Ms + N = 0$$

permits damped oscillatory approach to equilibration if the two roots are complex conjugates with negative real parts. Since these roots are

$$-M \pm \sqrt{(M^2 - N)}$$

the condition $M^2 - N < 0$ provides for oscillation at the angular frequency $\sqrt{(N - M^2)}$, amplitude-modulated by the decay e^{-Mt}.

Substitution of White's values above for e, f, g, and h in M and N shows that this set of values in fact produces oscillatory behaviour, which appropriately reverts to overdamped behaviour as f is stepped to smaller values. However, this striking agreement is entirely specious; the associated values of M are so large that the predicted oscillation would invariably be damped out in a fraction of a cycle and unmeasurable! Variations about the above values have not produced a more encouraging combination, and even White's (1973) values for the P_i response produce non-oscillatory behaviour. Although we cannot generalize about nonlinear systems of this sort, the conclusion from the above calculation is that either a very different kind of 3-state cycle is involved, or the basis of the isometric oscillation is to be sought elsewhere. (The latter is of course more attractive since White has shown remarkable matches between the three-state cycle and muscle.) This is potentially a rather strong and useful exclusion and ought to be pursued more formally.

APPENDIX III

NOTES ADDED MARCH 1974

Our review—largely a didactic presentation of muscle mechanics for the biochemist and of muscle biochemistry for the mechanist—went to press in *Progress in Biophysics and Molecular Biology* about a year ago. In updating it now for the present series, we have outlined some of the relevant developments of 1973 and 1974 in the notes below. Typographical errors in the original version, and a few slips of the pen,† have also been corrected.

1. *Myosin ATPase.* The kinetic scheme of Trentham and his coworkers for the hydrolysis of ATP by myosin subfragment 1 (summarized in formula (4a), §IV, 4) has been extended. The experimental basis of this work will not be discussed here, but can be found in two papers, by Bagshaw and Trentham (1973) and Bagshaw *et al.* (1974). The experiments have involved further analyses of the change in fluorescence upon rapidly mixing myosin and ATP. The conclusions from this work are that the binding of either ATP or ADP is a two-stage process, giving a revised form (with new nomenclature) of formula (4a):

$$\mathrm{M + ATP} \underset{k_{-1}}{\overset{k_1}{\rightleftarrows}} \mathrm{M{\cdot}ATP} \underset{k_{-2}}{\overset{k_2}{\rightleftarrows}} \mathrm{M^*ATP} \underset{k_{-3}}{\overset{k_3}{\rightleftarrows}} \mathrm{M^{**}ADP{\cdot}P} \underset{k_{-4}}{\overset{k_4}{\rightleftarrows}} \mathrm{M^*ADP{\cdot}P} \underset{k_{-5}}{\overset{k_5}{\rightleftarrows}}$$

$$\mathrm{M^*ADP + P} \underset{k_{-6}}{\overset{k_6}{\rightleftarrows}} \mathrm{M{\cdot}ADP} \underset{k_{\ 7}}{\overset{k_7}{\rightleftarrows}} \mathrm{M + ADP}$$

The product of the equilibrium constants for this reaction must (as pointed out by Taylor (1973) in his review of actin–myosin ATPase studies) be equal to the equilibrium constant for the hydrolysis of ATP (approximately 10^7) under the same ionic conditions. That is,

$$\prod_{i=1}^{7} K_i = 10^7.$$

It is possible (Taylor, personal communication) to specify certain products of equilibrium constants with a high degree of reliability. These are

(i) $K_1{\cdot}K_2 = 10^{13}$
(ii) $K_3{\cdot}K_4{\cdot}K_5 = 1$
(iii) $K_6{\cdot}K_7 = 10^{-6}$

Note that $K_1{\cdot}K_2{\cdot}K_3{\cdot}K_4{\cdot}K_5{\cdot}K_6{\cdot}K_7 = 10^7$ as required.

Oxygen isotope exchange studies (Bagshaw and Trentham, 1973) have established that reaction 3 in the above scheme is indeed a readily reversible cleavage step. A medium containing H_2O^{18} is found to exchange its oxygen readily with the terminal phosphate of the ATP in the M*ATP moiety.

Experiments by Mannherz *et al.* (1974) on the synthesis of ATP from ADP and P^{32} have indicated that the equilibrium constant for the rate-limiting step 4 in the above formula is about 12, and that the backward rate constant k_{-2} is exceedingly low (about 10^{-7} sec $^{-1}$). This implies that the step in the myosin ATPase cycle at which the greatest change in free energy occurs is step 2.

2. *Dual role of Ca^{++} in activation.* Our understanding of the control of activity has been extended by the finding that the basis of calcium sensitivity in many invertebrates is either a

†In particular, the A-filament spacings in Fig. 10 are, of course, 45 nm and 56 nm; the effective time constants of the Voigt element (§V, Fig. 27 and text) are $\eta_2/(k_1+k_2)$ and η_2/k_2 for length and tension steps, respectively (these were reversed in the original publication); the coefficients of the characteristic equation in Appendix II are expressed correctly in this edition.

light-chain subunit of the myosin subfragment 1 (in place of the tropomyosin–troponin interaction discussed in §IV, 2), or a dual system invoking both types of control. A good introduction to this subject is the paper by Lehman *et al.* (1972). See also the excellent review by Weber and Murray (1973).

3. *Role of phosphate.* In connexion with our finding (§VII, 2) that the mechanical transients obtained from *Lethocerus* are very different in the presence and absence of phosphate ions, it is of interest that a kinase which phosphorylates the light chains of myosin (Pires *et al.*, 1974) and another that phosphorylates two of the troponin molecules (Stull *et al.*, 1972; Perry and Cole, 1974) have recently been found. As yet, however, no significant effects associated with such phosphorylation have been discovered.

4. *Does glycerol extraction alter the contractile elements?* Material for mechanical experiments has included live fibres (Huxley and Simmons' work, some of Podolsky's and some of Julian's work), skinned fibres (live fibres from which the sarcolemma has been removed, used by Podolsky and his coworkers), and various types of extracted fibres, of which the most commonly used are glycerol-extracted; the latter have been used for most of the work on insect-flight muscle and rabbit psoas muscle. It would appear that there are a number of problems associated with the use of extracted fibres, so that one must be cautious in comparing the results of such experiments with those from experiments on live fibres. For example, (i) the Ca^{++} sensitivity of extracted fibres changes with time (see, for example, Abbott, 1973), (ii) the myokinase activity changes with time (Abbott and Leech, 1973), and (iii) the ability of glycerol-extracted fibres to develop rigor tension when the ATP concentration is lowered in the absence of Ca^{++} (White, 1970) suggests that the myosin-bound Ca^{++} sensitivity, conferred by one of the subunits of myosin subfragment 1, is lost (Lehman, Bullard and Hammond, 1974). In connexion with this last point we suggest that one criterion defining a suitable extraction medium for muscles containing myosin-bound Ca^{++} sensitivity may be the inability to develop rigor tension.

5. *Contents of experimental solutions.* We neglected to mention a valuable yet little known procedure for estimating the equilibrium concentration of ionic and complex species in solutions prepared as bathing media. Since the equilibrium equations for the concentrations of n species usually involve n simultaneous nonlinear algebraic equations, it is usual to estimate the expected concentrations in a rough fashion only, or merely to state the recipe for preparation of the medium. Perrin and Sayce (1967) have applied a computer algorithm appropriate for the form of such equations and published their FORTRAN program for it. See, for example, White and Thorson (1972), in which the association constants and equilibrium concentrations of thirty-seven complex species used in muscle studies are treated.

6. *Rapid mechanical transients.* Huxley and Simmons have improved the apparatus described in §V so that length changes can now be applied in 0.2 msec and the response measured via a tension transducer with a resonant frequency of over 10 kHz. Data obtained with the use of this improved apparatus (Ford, Huxley and Simmons, 1974) support their earlier conclusions (§VI, 3).

7. *Thermodynamic constraints* upon the rate constants of the cross-bridge cycle (§VI, 6) are now receiving increased attention. See, for example, T.L. Hill's (1974) forthcoming article and A. Cooper's (1974) analysis of the energetics of the contractile cycle.

References to Appendix III

ABBOTT, R. H. (1973) The effects of fibre length and calcium ion concentration on the dynamic response of glycerol-extracted insect fibrillar muscle, *J. Physiol.* **231,** 195–208.

ABBOTT, R. H. and LEECH, A. R. (1973) Persistence of adenylate kinase and other enzymes in glycerol-extracted muscle, *Pflügers Archiv* **344,** 233–43.

BAGSHAW, C. R. and TRENTHAM, D. R. (1973) The reversibility of adenosine triphosphate cleavage by myosin, *Biochem. J.* **133,** 323–8.

BAGSHAW, C. R., ECCLESTON, J. F., ECKSTEIN, F., GOODY, R. S., GUTFREUND, H. and TRENTHAM, D. R. (1974) The magnesium ion-dependent adenosine triphosphatase of myosin. *Biochem. J.* **141,** 351–64.

COOPER, A. (1974) An analysis of the possible source of contractile forces in striated muscle, *J. theoret. Biol.* **42,** 545–62.

FORD, L., HUXLEY, A. F. and SIMMONS, R. M. (1974) *J. Physiol.* (Proceedings of the Physiological Society). In press.

HILL, T. L. (1974) Theoretical formalism for the sliding filament model of contraction of striated muscle. *Progress Biophys. Mol. Biol.* **28** (in press).

HITCHCOCK, S. E., HUXLEY, H. E. and SZENT-GYÖRGYI, A. G. (1973) Calcium sensitive binding of troponin to actin-tropomyosin: a two-site model for troponin action. *J. mol. Biol.* **80,** 825–36.

HUXLEY, A. F. (1973) A note suggesting that the cross-bridge attachment during muscle contraction may take place in two stages, *Proc. Roy. Soc. B,* **183,** 83–6.

KENDRICK-JONES, J., LEHMAN, W. and SZENT-GYÖRGYI, A. G. (1970) Regulation in molluscan muscles, *J. mol. Biol.* **54,** 313–26.

LEHMAN, W., KENDRICK-JONES, J. and SZENT-GYÖRGYI, A. G. (1972) Myosin linked regulatory systems: Comparative studies, *Cold Spring Harbor Symp. Quant. Biol.* **37,** 319–30.

LEHMAN, W. and SZENT-GYÖRGYI, A. G. (1972) Activation of the adenosine triphosphatase of *Limulus polyphemus* actomyosin by tropomyosin, *J. gen. Physiol.* **59,** 375–87.

LEHMAN, W., BULLARD, B. and HAMMOND, K. (1974) Calcium sensitive myosin in insect flight muscles (in preparation).

LYMN, R. W. (1974) Effect of modifier on three-step enzymic process: general rate equation, *J. theoret. Biol.* **43,** 305–12.

LYMN, R. W. (1974) Actin activation of the myosin ATPase: A kinetic analysis, *J. theoret. Biol.* **43,** 313–28.

MANNHERZ, H. G., SCHENCK, H. and GOODY, R. S. (1974) Synthesis of ATP from ADP and P_i at the myosin-S1 active site (in preparation).

MARSTON, S. (1973) The nucleotide complexes of myosin in glycerol-extracted muscle fibres. *Biochim. Biophys. Acta* **305,** 397–412.

MURRAY, J. M. and WEBER, A. (1974) The cooperative action of muscle proteins, *Sci. Amer.* **230** (2), 69–71.

PERRIE, W. T., SMILLIE, L. B. and PERRY, S. V. (1973) A phosphorylated light-chain component of myosin from skeletal muscle, *Biochem. J.* **135,** 151–64.

PERRIN, D. D. and SAYCE, I. G. (1967) Computer calculation of equilibrium concentrations in mixtures of metal ions and complexing species, *Talanta,* **14,** 833.

PERRY, S. V. and COLE, H. A. (1973) Phosphorylation of the "3700 component" of the troponin complex (Troponin-T), *Biochem. J.* **131,** 425–8.

PERRY, S. V. and COLE, H. A. (1974) *Biochem. J.* (in press).

PIRES, E., PERRY, S. V. and THOMAS, M. A. W. (1974) Myosin light chain kinase; a new enzyme from striated muscle. *F.E.B.S. Letters* **41,** 292–6.

STULL, J. T., ENGLAND, P. J., BROSTROM, C. O. and KREBS, E. G. (1972) Phosphorylation and dephosphorylation of the inhibitory component of troponin, *Cold Spring Harbor Symp. Quant. Biol.* **37,** 263–6.

SZENT-GYÖRGYI, A. G., SZENTKIRÁLYI, E. M. and KENDRICK-JONES, J. (1973) The light chains of scallop myosin as regulatory subunits, *J. mol. Biol.* **74,** 179–203.

TAYLOR, E. W. (1973) Mechanism of actomyosin ATPase and the problem of muscle contraction, *Current Topics in Bioenergetics,* **5,** 201–31.

WEBER, A. and MURRAY, J. M. (1973) Molecular control mechanisms in muscle contraction, *Physiol. Revs.* **53,** 612–73.

ACKNOWLEDGEMENTS

We wish to thank Dr. Ann Biederman-Thorson and Miss Madelaine Green for considerable help and advice in the preparation of the paper. Professors J. D. Currey, A. F. Huxley, Drs. R. H. Abbott, E. Eisenberg, G. Kellett, A. Miller, G. Penzer, R. J. Podolsky, R. M. Simmons, E. W. Taylor, and D. R. Trentham kindly read parts of or the complete manuscript, and made very helpful comments. The following authors of papers from the 37th Cold Spring Harbor Symposium of Quantitative Biology generously allowed us to read their papers before publication: Abbott; Armitage, Miller, Rodger, and Tregear; Bagshaw, Eccleston, Trentham, and Yates; Eisenberg and Kielley; Huxley and Simmons; Julian, Sollins, and Sollins; Podolsky and Nolan; Pybus and Tregear.

REFERENCES

ABBOTT, R. H. (1968) The mechanism of oscillatory contraction of insect fibrillar flight muscle, D.Phil. thesis, University of Oxford.

ABBOTT, R. H. (1972) Comments on the mechanism of force generation in striated muscles, *Nature New Biol.* **239,** 183–6.

ABBOTT, R. H. (1973) Does calcium affect the rate constants of muscle? *37th Cold Spring Harbor Symposium of Quantitative Biology.*

AIDLEY, D. J. (1971) *The Physiology of Excitable Cells*, Cambridge University Press.

AIDLEY, D. J. and WHITE, D. C. S. (1969) Mechanical properties of glycerinated fibres from the Tymbal muscles of Brazilian Cicada, *J. Physiol.* **205,** 179–92.

ARMITAGE, P., MILLER, A., RODGER, C. D., and TREGEAR, R. T. (1973) Structure and function of insect flight muscle, *37th Cold Spring Harbor Symposium of Quantitative Biology.*

ARMSTRONG, C. F., HUXLEY, A. F., and JULIAN, F. J. (1966) Oscillatory responses in frog skeletal muscle fibres, *J. Physiol.* **186,** 26–7P.

ASAKURA, S., TANIGUCHI, M., and OOSAWA, F. (1963) Mechano-chemical behaviour of F-actin, *J. Molec. Biol.* **7,** 55–69.

ASHHURST, D. E. (1971) The *Z*-line in insect flight muscle, *J. Molec. Biol.* **55,** 283–5.

AUBER, J. and COUTEAUX, R. (1963) Ultrastructure de la strie Z dans des muscles de diptères, *J. de Microscopie* **2,** 309–24

BAGSHAW, C. R., ECCLESTON, J. F., TRENTHAM, D. R., and YATES, D. W. (1973) Transient kinetic studies of the magnesium-dependent ATP-ase of myosin and its proteolytic subfragments, *37th Cold Spring Harbor Symposium of Quantitative Biology.*

BARANY, M. and FINKLEMAN, F. (1962) The lability of the F-actin bound calcium under ultrasonic vibration, *Biochim. biophys. Acta* **63,** 98–105.

BARANY, M. and FINKLEMAN, F. (1963) The exchange of the F-actin-bound ADP, *Biochim. biophys. Acta* **78,** 175–95.

BERNHARD, J. A. (1968) *The Structures and Function of Enzyme*, Benjamin.

BLIX, M. (1893) Die Länge und die Spannung des Muskels, *Skand. Arch. Physiol.* **4,** 399–409.

BREMEL, R. D. and WEBER, A. (1972) Cooperation within the actin filament in vertebrate skeletal muscle, *Nature New Biol.* **238,** 97–101.

BULLARD, B., DABROWSKA, R., and WINKLEMAN, L. (1972) Hybrid actomyosins, *Abstracts of 4th International Congress Biophysics and Biochemistry* (*Moscow*) and (1973) *Biochem. J.* **135,** 277–86.

BULLARD, B., LUKE, B., and WINKLEMAN, L. (1973) The paramyosin of insect flight muscle, *J. Mol. Biol.* **75,** 359–67.

BULLARD, B. and REEDY, M. K. (1973) How many myosins per cross-bridge, II. 37th *Cold Spring Harbor Symposium of Quantitative Biology.*

CARLSON, F. D., BONNER, R. F., and FRASER, A. (1973) Intensity fluctuation spectra from resting and contracting muscle, *37th Cold Spring Harbor Symposium of Quantitative Biology.*

CASPAR, D. L. D., COHEN, C., and LONGLEY, W. (1969) Tropomyosin: crystal structure, polymorphism and molecular interactions, *J. Molec. Biol.* **41,** 87–107.

CHAPLAIN, R. A. and TREGEAR, R. T. (1966) The mass of myosin per cross-bridge in insect fibrillar flight muscle, *J. Molec. Biol.* **21,** 275–80.

CIVAN, M. M. and PODOLSKY, R. J. (1966) Contraction kinetics of striated muscle fibres following quick changes in load, *J. Physiol.* **184,** 511–34.

COHEN, C., CASPAR, D. L. D., PARRY, D. A. D., and LUCAS, R. M. (1971) Tropomyosin crystal dynamics, *Cold Spring Harbor Symposium of Quantitative Biology* **36,** 205–16.

CURTIN, N. A. and DAVIES, R. E. (1973) Mechanical and chemical properties of muscle during stretching, *37th Cold Spring Harbor Symposium of Quantitative Biology.*

DAVIES, R. E. (1964) Adenosine triphosphate breakdown during single muscle contractions, *Proc. Roy. Soc.* B, **160,** 480–4.

DESCHEREVSKY, V. I. (1968) *Biophysica* **13,** 928.

DOS REMEDIOS, C. G., MILLIKAN, R. G. C., and MORALES, M. F. (1972) Polarization of tryptophan fluorescence from single striated muscle fibres, *J. Gen. Physiol.* **59,** 103–20.

EBASHI, S., OHTSUKI, I. and MIHASHI, K. (1973) Regulatory proteins of muscle with special reference to troponin, *37th Cold Spring Harbor Symposium of Quantitative Biology.*

EISENBERG, E., DOBKIN, L., and KIELLEY, W. W. (1972) Heavy meromyosin: evidence for a refractory state unable to bind to actin in the presence of ATP, *Proc. Natn. Acad. Sci. USA* **69,** 667–71.

EISENBERG, E. and KIELLEY, W. W. (1973) Evidence for a refractory state of heavy meromyosin and subunit-1 unable to bind to actin in the presence of ATP, *37th Cold Spring Harbor Symposium of Quantitative Biology.*

EISENBERG, E. and MOOS, C. (1967) The interaction of actin with myosin and heavy meromyosin in solution at low ionic strength, *J. Biol. Chem.*, **242,** 2945–51.

EISENBERG, E. and MOOS, C. (1968) Adenosine triphosphate activity of acto-heavy meromyosin: a kinetic analysis of actin activation. *Biochemistry* **7,** 1486–9.

EISENBERG, E. and MOOS, C. (1970) Actin activation of heavy-meromyosin adenosine triphosphatase, *J. Biol. Chem.* **245,** 2451–6.

ELLIOTT, G. F., LOWY, J., and WORTHINGTON, C. R. (1963) An X-ray and light diffraction study of the filament lattice of striated muscle in the living state and rigor, *J. Molec. Biol.* **6,** 295–305.

ELZINGA, M. and COLLINS, J. H. (1973) The amino-acid sequence of rabbit skeletal muscle actin, *37th Cold Spring Harbor Symposium of Quantitative Biology.*

FENN, W. O. (1924) A quantitative comparison between the energy liberated and the work performed by the isolated sartorius muscle of the frog, *J. Physiol.* **58,** 175–203.

FENN, W. C. and MARSH, B. S. (1935) Muscular force at different speeds of shortening, *J. Physiol.* **85,** 277–97.

FINLAYSON, B., LYMN, R. W., and TAYLOR, E. W. (1969) Studies on the kinetics of formation and dissociation of the actomyosin complex, *Biochemistry* **8,** 811–19.

GARAMVOLGYI, N. (1969) The structural basis of the elastic properties in the flight muscle of the bee, *J. Ultrastruct. Res.* **27,** 462–71.

GASSER, H. S. and HILL, A. V. (1924) The dynamics of muscular contraction, *Proc. Roy. Soc.* B, **96,** 398–437.

GOODALL, M. C. (1956) Auto-oscillations in extracted muscle fibre systems, *Nature* **177,** 1238–9.

GORDON, A. M., HUXLEY, A. F., and JULIAN, F. J. (1966) The variation in isometric tension with sarcomere length in vertebrate muscle fibres, *J. Physiol.* **184,** 170–92.

GREASER, M. L. and GERGELY, J. (1973) Troponin subunits and their interactions, *37th Cold Spring Harbor Symposium of Quantitative Biology.*

GUTFREUND, H. (1965) *An Introduction to the Study of Enzymes*, Blackwell Scientific Publications.

HANSON, J. and HUXLEY, H. E. (1957) Quantitative studies on the structure of cross-striated myofibrils: II, Investigations by biochemical techniques, *Biochim. biophys. Acta* **23,** 250–60.

HANSON, J. and LOWY, J. (1963) The structure of F-actin and of actin filaments isolated from muscle, *J. Molec. Biol.* **6,** 46–60.

HARTSHORNE, D. J. (1973) Studies on the subunit composition of troponin, *37th Cold Spring Harbor Symposium of Quantitative Biology.*

HEINL, P. (1972) Mechanische Activierung und Deaktivierung der isolierten contractilen Struktur des Froschsartorius durch rechteckförmige und sinusförmige Längenänderungen, **333,** 213–25.

HILL, A. V. (1922) The maximal work and mechanical efficiency of human muscles, *J. Physiol.* **56,** 19–41.

HILL, A. V. (1938) Methods of analysing the heat production of muscle, *Proc. Roy. Soc.* B, **124,** 114–36.

HILL, D. K. (1968) Tension due to interaction between the sliding filaments in resting striated muscle: the effect of stimulation, *J. Physiol.* **19,** 637–84.

HILL, T. L. (1966) Diagrammatic representation of steady-state fluxes for unmolecular systems, *J. Theor. Biol.* **10,** 442–59.

HILL, T. L. (1968a) Phase transition in the sliding filament model of muscle contraction, *Proc. Natn. Acad. Sci. USA* **59,** 1194–1200.

HILL, T. L. (1968b) On the sliding filament model of muscular contraction, *Proc. Natn. Acad. Sci. USA* **61,** 98–105.

HILL, T. L. and WHITE, G. M. (1968a) Kinetics of cross-bridge fluctuations in configuration, *Proc. Natn. Acad. Sci. USA* **61,** 514–21.

HILL, T. L. and WHITE, G. M. (1968b) Calculation of force–velocity curves, *Proc. Natn. Acad. Sci. USA* **61,** 889–96.

HUXLEY, A. F. (1957) Muscle structure and theories of contraction, *Prog. Biophys.* **7,** 255–318.

HUXLEY, A. F. (1971) The Croonian Lecture: the activation of striated muscle and its mechanical response. *Proc. Roy. Soc.* B, **178,** 1–27.

HUXLEY, A. F. and NIEDERGERKE, R. (1954) Structural changes in muscle during contraction. *Nature* **173,** 971–3.

HUXLEY, A. F. and SIMMONS, R. M. (1968) A capacitance-gauge tension transducer, *J. Physiol.* **197,** 12P.

HUXLEY, A. F. and SIMMONS, R. M. (1971a) Mechanical properties of the cross bridges of frog and striated muscle. *J. Physiol.* **218,** 59–60P.

HUXLEY, A. F. and SIMMONS, R. M. (1971b) Proposed mechanism of force generation in striated muscle. *Nature* **233,** 533–8.

HUXLEY, A. F. and SIMMONS, R. M. (1972) Reply to Abbott. *Nature New Biol.* **239,** 186–7.

HUXLEY, A. F. and SIMMONS, R. M. (1973) Mechanical transients and the origin of muscular force, *37th Cold Spring Harbor Symposium of Quantitative Biology.*

HUXLEY, H. E. (1953) Electron microscope studies on the organisation of the filaments in striated muscle, *Biochim. biophys. Acta* **12,** 387–94.

HUXLEY, H. E. (1957) The double array of filaments in cross-striated muscle, *J. Biophys. Biochem. Cytol.* **3,** 631–47.

HUXLEY, H. E. (1963) Electron microscope studies on the structure of natural and synthetic protein filaments from striated muscle, *J. Molec. Biol.* **7,** 281–308.

HUXLEY, H. E. (1969) The mechanism of muscle contraction, *Science* **164,** 1356–66.

HUXLEY, H. E. (1971) The Croonian Lecture, 1970: the structural basis of muscular contraction, *Proc. Roy, Soc.* B, **178,** 131–49.

HUXLEY, H. E. (1973) Factors controlling the movement and attachment of the cross-bridges in muscle, *37th Cold Spring Harbor Symposium of Quantitative Biology.*

HUXLEY, H. E. and BROWN, W. (1967) The low angle X-ray diagram of vertebrate striated muscle and its behaviour during contraction and rigor, *J. Molec. Biol.* **30,** 383–434.

HUXLEY, H. E. and HANSON, J. (1954) Changes in the cross-striations of muscle during contraction stretch and their structural interpretation, *Nature* **173,** 973–6.

HUXLEY, H. E. and HANSON, J. (1957) Quantitative studies on the structure of cross-striated myofibrils: I, Investigations by interference microscopy, *Biochim. biophys. Acta* **23,** 229–49. II, Investigations by biochemical techniques, *Biochim. biophys. Acta* **23,** 250–60.

INOUE, A., SHIBATA-SEKIYA, K., and TONOMURA, Y. (1972) The pre-steady state of the myosin-adenosine triphosphatase system: XI, Formation and decomposition of the reactive myosin-phosphate-ADP complex, *Jap. J. Biochem.* **71,** 115–24.

JEWELL, B. R. and RÜEGG, C. (1966) Oscillatory contraction of insect fibrillar muscle after glyceral extraction, *Proc. Roy. Soc.* B, **164,** 428–59.

JEWELL, B. R. and WILKIE, D. R. (1958) An analysis of the mechanical components in frog's striated muscle, *J. Physiol.* **143,** 515–40.

JULIAN, F. (1969) Activation in a skeletal muscle contraction model with a modification of insect fibrillar muscle, *Biophys. J.* **9,** 547–70.

JULIAN, F., SOLLINS, K. R., and SOLLINS, M. R. (1973) A model for muscle contraction in which cross-bridge attachment and force generation are distinct, *37th Cold Spring Harbor Symposium of Quantitative Biology.*

KUSHMERICK, M., LARSON, R. E., and DAVIES, R. E. (1969) The chemical energetics of muscle contraction: 1, Activation heat, heat of shortening and ATP utilisation for activation-relaxation processes, *Proc. Roy. Soc.* B, **174,** 293–313.

LEADBETTER, L. and PERRY, S. V. (1963) The effect of actin on magnesium-activated adenosine triphosphatase of heavy meromyosin, *Biochem. J.* **87,** 233–8.

LEVIN, A. and WYMAN, J. (1927) The viscous elastic properties of muscle, *Proc. Roy. Soc.* B, **101,** 218–43.

LORAND, L. and MOOS, C. (1956) Auto-oscillations in extracted muscle fibre systems, *Nature* **177,** 1239.

LOWEY, S., GOLDSTEIN, L., COHEN, C., and LUCK, S. M. (1966) Proteolytic degradation of myosin and the meromyosins by a water-insoluble polyanionic derivative of trypsin, *J. Molec. Biol.* **23,** 287–304.

LOWEY, S. and LUCK, S. M. (1969) Equilibrium binding of adenosine diphosphate to myosin. *Biochemistry* **8,** 3195–9.

LYMN, R. W. and TAYLOR, E. W. (1970) Transient state phosphate production in the hydrolysis of nucleoside triphosphates by myosin, *Biochemistry* **9,** 2975–83.

LYMN, R. W. and TAYLOR, E. W. (1971) The mechanism of adenosine triphosphate hydrolysis by actomyosin, *Biochemistry* **10,** 4617–24.

MACHIN, K. E. and PRINGLE, J. W. S. (1959) The physiology of insect flight muscle: II, The mechanical properties of a beetle flight muscle, *Proc. Roy. Soc.* B, **151,** 204–25.

MACHIN, K. E. and PRINGLE, J. W. S. (1960) The physiology of insect fibrillar muscle: III, The effect of sinusoidal changes of length on a beetle flight muscle, *Proc. Roy. Soc.* B, **152,** 311–30.

MARSTON, S. B. and TREGEAR, R. T. (1972) Evidence for a complex between myosin and ADP in relaxed muscle fibres, *Nature New Biol.* **235,** 23–4.

MARUYAMA, K. and GERGELY, J. (1962) Interaction of actomyosin with ATP at low ionic strength: II, Factors influencing clearing and superprecipitation. ATPase and birefringence of flow studies, *J. Biol. Chem.* **237,** 1100–6.

MILLER, A. and TREGEAR, R. T. (1972) The structure of insect flight muscle in the presence and absence of ATP, *J. Molec. Biol.* **70,** 85–104.

MOORE, P. B., HUXLEY, H. E., and DEROSIER, D. J. (1970) 3-D reconstruction of F-actin, thin filaments and decorated thin filaments, *J. Molec. Biol.* **50,** 279–95.

MOOS, C. (1973) Actin activation of heavy meromyosin and subunit-1 ATPases: steady state kinetics studies, *37th Cold Spring Harbor Symposium of Quantitative Biology.*

MORALES, M. F. (1959) The mechanism of muscle contraction, in *Biophysical Science—A Study Program* (ed. J. L. Oncley), pp. 426–32.

MURPHY, G. M. (1960) *Ordinary Differential Equations and their Solution,* van Nostrand Reinhold Co., New York.

OFFER, G. W. (1973) C-protein and the periodicity in the thick filament assembly, *37th Cold Spring Harbor Symposium of Quantitative Biology.*

PAGE, S. G. and HUXLEY, H. E. (1963) Filament lengths in striated muscle, *J. Molec. Biol.* **9,** 369–90.

PEPE, F. A. (1966) Some aspects of the structural organisation of the myofibril as revealed by antibody staining methods, *J. Cell Biol.* **28,** 505–25.

PEPE, F. A. (1973) The myosin filament, *37th Cold Spring Harbor Symposium of Quantitative Biology.*

PERRY, S. V., COLE, H. A., HEAD, J. F., and WILSON, F. J. (1973) Localisation and mode of action of the inhibitory protein component of the troponin complex, *37th Cold Spring Harbor Symposium of Quantitative Biology.*

PLATT, J. R. (1964) Strong inference, *Science* **146,** 347–52.

PODOLSKY, R. J. (1960) Kinetics of muscular contraction: the approach to the steady state, *Nature* **188,** 666–8.

PODOLSKY, R. J. and NOLAN, A. C. (1972) Cross-bridge properties derived from physiological studies of frog muscle fibres, in *Contractility of Muscle Cells and Related Processes* (ed. R. J. Podolsky), pp. 247–60.

PODOLSKY, R. J. and NOLAN, A. C. (1973) Muscle contraction transients, cross-bridge kinetics and the Fenn Effect, *37th Cold Spring Harbor Symposium of Quantitative Biology.*

PODOLSKY, R. J., NOLAN, A. C., and ZAVELIER, S. A. (1969) Cross-bridge properties derived from muscle isotonic velocity transients, *Proc. Natn. Acad. Sci. USA* **64,** 504–11.

POLISSAR, M. J. (1952) Physical chemistry of contractile process in muscle: I, A physicochemical model of the contractile mechanism, *Am. J. Physiol.* **168,** 766–805.

PRINGLE, J. W. S. (1949) The excitation and contraction of the flight muscles of insects, *J. Physiol.* **108,** 226–32.

PRINGLE, J. W. S. (1960) Models of muscle, *Symp. Soc. Exp. Biol.* **14,** 41–68.

PRINGLE, J. W. S. (1967) The contractile mechanism of insect fibrillar muscle, *Prog. Biophys.* **17,** 1–60.

PRINGLE, J. W. S. and TREGEAR, R. T. (1969) Mechanical properties of insect fibrillar muscle at large amplitudes of oscillation, *Proc. Roy. Soc.* B, **174,** 33–50.

PYBUS, J. (1972) ATP hydrolysis by muscle and related topics, D.Phil thesis, University of Oxford.

PYBUS, J. and TREGEAR, R. T. (1973) Estimates of how long and how hard myosin heads pull on actin and how many do so at one time, *37th Cold Spring Harbor Symposium of Quantitative Biology.*

RAMSEY, R. W. and STREET, S. F. (1940) Isometric length–tension diagrams of isolated skeletal muscle fibres of the frog, *J. Cell Comp. Physiol.* **15,** 11–34.

REEDY, M. K. (1968) Ultrastructure of insect flight muscle: I, Screw sense and structural grouping in the rigor cross-bridge lattice, *J. Molec. Biol.* **31,** 155–76.

REEDY, M. K. (1972) Electron microscope observations concerning the behaviour of the cross-bridge in striated muscle, in *Contractility of Muscle Cells and Related Processes* (ed. R. J. Podolsky), pp. 229–46.

REEDY, M. K., HOLMES, K. C., and TREGEAR, R. T. (1965) Induced changes in orientation of the cross-bridges of glycerated insect flight muscle, *Nature* **207,** 1276–80.

REEDY, M. K., BAHR, G. F. and FISCHMAN, D. A. (1973) How many myosins per cross-bridge, I. *37th Cold Spring Harbor Symposium of Quantitative Biology.*

RIZZINO, A. A., BAROUCH, W. W., EISENBERG, E., and MOOS, C. (1970) Actin-heavy meromyosin binding: determination of binding stoichiometry from ATPase kinetic measurements, *Biochemistry* **9,** 2402–7.

RÜEGG, J. C., STEIGER, G. J., and SCHÄDLER, M. (1970) The mechanical activation of the contractile system on skeletal muscle, *Pflügers Arch.* **319,** 139–45.

RÜEGG, J. C., SCHÄDLER, M., STEIGER, G. J., and MULLER, G. (1971) Effects of inorganic phosphate on the contractile mechanism, *Pflügers Arch.* **325,** 359–64.

RÜEGG, J. C. and STUMPF, H. (1969a) The coupling of power output and myo-fibrillar ATPase activity in glycerol-extracted insect fibrillar muscle at varying amplitude of ATP-driven oscillation, *Pflügers Arch.* **305,** 21–33.

RÜEGG, J. C. and STUMPF, H. (1969b) Activation of the myofibrillar ATPase activity by extension of glycerol-extracted insect fibrillar muscle, *Pflügers Arch.* **305,** 34–46.

RÜEGG, J. C. and TREGEAR, R. T. (1966) Mechanical factors affecting the ATPase activity of glycerol-extracted insect fibrillar flight muscle, *Proc. Roy. Soc.* B, **165,** 497–512.

SCHÄDLER, M., STEIGER, G., and RÜEGG, J. C. (1969) Tension transients in glycerol-extracted fibres of insect fibrillar muscle, *Experientia* **25,** 942–3.

SCHÄDLER, M., STEIGER, G. J., and RÜEGG, J. C. (1971) Mechanical activation and isometric oscillation in insect fibrillar muscle, *Pflügers Arch.* **330,** 217–29.

SIMMONS, R. M. and JEWELL, B. R. (1973) Mechanics and models of muscular contraction, in: *Recent Advances in Physiology*. ed. R. J. Linden. J. and A. Churchill Ltd.

SMITH, D. S. (1966) The organisation and function of the sarcoplasmic reticulum and T-system of muscle cells, *Prog. Biophys.* **16,** 107–42.

SQUIRE, J. M. (1971) General model for the structure of all myosin-containing filaments, *Nature* **223,** 457–62.

STEIGER, G. J. and RÜEGG, J. C. (1969) Energetics and "efficiency" in the isolated contractile machinery of an insect fibrillar muscle at various frequencies of oscillation, *Pflügers Arch.* **307,** 1–21.

STEVENS, H. C. and METCALF, R. P. (1934) The decrement of muscular force with increasing speed of shortening, *Am. J. Physiol.* **107,** 568–76.

STROMER, M. H. and GOLL, D. E. (1972) Studies on purified α-actinin: II, Electron microscope studies on the competitive binding of α-actinin and tropomyosin to *Z* line extracted myofibrils, *J. Molec. Biol.* **67,** 489–94.

SZENT-GYORGYI, A. G. and PRIOR, G. (1966) Exchange of ADP bound to actin in superprecipitated actomyosin and contracted myofibrils, *J. Molec. Biol.* **15,** 515–38.

SZENT-GYORGYI, A. G., COHEN, C., and KENDRICK-JONES (1971) Paramyosin and the filaments of molluscan "catch" muscles, *J. Molec. Biol.* **56,** 239–58.

THORSON, J. and BIEDERMAN-THORSON, M. (1974) Distributed relaxation processes in sensory adaptation, *Science* **183,** 161–72.

THORSON, J. W. and WHITE, D. C. S. (1969) Distributed representations for actin-myosin interaction in the oscillatory contraction of muscle, *Biophys. J.* **9,** 360–90.

TREGEAR, R. T. (1973) The biophysics of Insect Muscle. in: *Advances in Insect Physiology*, ed. P. N. R. Usherwood. In press.

TREGEAR, R. T. and MILLER, A. (1969) Evidence of cross-bridge movement during contraction of insect flight muscle, *Nature* **222,** 1184–85.

TRENTHAM, D. R., BARDSLEY, R. G., ECCLESTON, J. F. and WEEDS, A. G. (1972) Elementary processes of the magnesium ion-dependent ATPase activity of heavy meromyosin, *Biochem. J.* **126,** 635–44.

WEBER, E. (1846) Muskelbewegung, *Handworterbuch der Physiol.*, p.1.

WHITE, D. C. S. (1967) Structural and mechanical properties of insect fibrillar flight muscle in the relaxed and rigor states, D.Phil. thesis, University of Oxford.

WHITE, D. C. S. (1970) Rigor contraction and the effect of various phosphate compounds on glycerinated insect flight and vertebrate muscle, *J. Physiol.* **208,** 583–605.

WHITE, D. C. S. (1973) Dynamics of contraction in insect flight muscle, *37th Cold Spring Harbor Symposium of Quantitative Biology.*

WHITE, D. C. S. and THORSON, J. T. (1972) Phosphate starvation and the nonlinear dynamics of insect fibrillar flight muscle, *J. Gen. Physiol.* **60,** 307–36.

YOUNG, D. M. (1967) Studies on the structural basis of the interaction between myosin and actin, *Proc. Natn. Acad. Sci. USA* **58,** 2393–400.

ZEBE, E., MEINRENKEN, W., and RÜEGG, J. C. (1968) Superkontraktion glyzerinextrahierter asynchroner Insektenmuskeln in Gegenwart von ITP, *Z. Zellforsch.* **87,** 603–21.

INDEX